D1104095

Contesting the Future of
Nuclear Power
A Critical Global Assessment of Atomic Energy

Contesting the Future of
Nuclear Power

A Critical Global Assessment of Atomic Energy

Benjamin K Sovacool

National University of Singapore, Singapore

World Scientific

NEW JERSEY · LONDON · SINGAPORE · BEIJING · SHANGHAI · HONG KONG · TAIPEI · CHENNAI

Published by

World Scientific Publishing Co. Pte. Ltd.

5 Toh Tuck Link, Singapore 596224

USA office: 27 Warren Street, Suite 401-402, Hackensack, NJ 07601

UK office: 57 Shelton Street, Covent Garden, London WC2H 9HE

British Library Cataloguing-in-Publication Data
A catalogue record for this book is available from the British Library.

CONTESTING THE FUTURE OF NUCLEAR POWER
A Critical Global Assessment of Atomic Energy

ISBN-13 978-981-4322-75-1
ISBN-10 981-4322-75-X

Typeset by Stallion Press
Email: enquiries@stallionpress.com

Printed in Singapore by World Scientific Printers.

Acknowledgments

The author is appreciative to participants from the "2009 EnviroEnergy International Conference on Energy and Environment" held at the Taj Chandigarh Hotel in Chandigarh, India; the "World Renewable Energy Congress Asia 2009" held at the Bangkok International Trade and Exhibition Centre in Thailand; the "Nuclear Politics, Policy, and Planning Workshop" of the 2009 Nordic Environment Social Sciences Conference held at the University College London Environment Institute in London, United Kingdom; the 2009 "Managing Radioactive Waste: Problems and Challenges in a Globalizing World" Conference held at the University of Gothenburg, Sweden; the 51st Annual Convention of the International Studies Association held in New Orleans, United States; and the 7th Transatlantic Energy Governance Dialogue, "Toward a Nuclear Power Renaissance? Challenges for Global Energy Governance," held at the Hotel Griebnitzsee in Potsdam, Germany, hosted by the Global Public Policy Institute and the Brookings Institution, for helpful comments on some of the chapters in this book. Christopher Cooper from the Vermont Law School, whose energy policy insight and general sexiness remain unrivaled, helped with conceptualizing many of the arguments presented in the book. Anthony D'Agostino provided much-needed help in researching some of the claims advanced within the book as well. The author is extremely grateful to Antony Froggatt from Chatham House, Sharon Squassoni from the Center for Strategic and International Studies, Andreas

Goldthau from the Global Public Policy Institute, Wolfgang Dirschauer from Vattenfall, Mark E. Gaffigan from the US Government Accountability Office, Cristoph Pistner from the Institute for Applied Ecology, and Steve Curtis from the UNLV's Harry Reid Center for providing critical comments and suggestions on earlier drafts of various chapters. Mark A. Delucchi from the University of California (UC) Davis, Paul Denholm from the National Renewable Energy Laboratory, Roberto Dones from the Swiss Laboratory for Energy Systems Analysis, V.M. Fthenakis from Brookhaven National Laboratory, Paul J. Meier from the University of Wisconsin–Madison, and Jan Willem Storm van Leeuwen provided invaluable and outstanding comments and suggestions for the climate change and carbon footprint sections of this book. All errors and conclusions, however, remain solely my own.

Lastly, portions of the book draw from arguments and data published previously in Benjamin K. Sovacool, "Think Again: Nuclear Power," *Foreign Policy* 150 (September/October, 2005), pp. 1–4; Benjamin K. Sovacool, "Coal and Nuclear Technologies: Creating a False Dichotomy for American Energy Policy," *Policy Sciences* 40(2) (June, 2007), pp. 101–122; Benjamin K. Sovacool, "The Costs of Failure: A Preliminary Assessment of Major Energy Accidents, 1907 to 2007," *Energy Policy* 36(5) (May, 2008), pp. 1802–1820; Benjamin K. Sovacool, "Valuing the Greenhouse Gas Emissions from Nuclear Power: A Critical Survey," *Energy Policy* 36(8) (August, 2008), pp. 2940–2953; Benjamin K. Sovacool and Christopher Cooper, "Nuclear Nonsense: Why Nuclear Power Is No Answer to Climate Change and the World's Post-Kyoto Energy Challenges," *William & Mary Environmental Law and Policy Review* 33(1) (Fall, 2008), pp. 1–119; Benjamin K. Sovacool, "Running on Empty: The Electricity–Water Nexus and the U.S. Electric Utility Sector," *Energy Law Journal* 30(1) (April, 2009), pp. 11–51; Benjamin K. Sovacool and Charmaine Watts, "Going Completely Renewable: Is It Possible (Let Alone Desirable)?," *Electricity Journal* 22(4) (May, 2009), pp. 95–111; Benjamin K. Sovacool and Kelly Sovacool, "Identifying Future Electricity–Water Tradeoffs in the United States," *Energy Policy* 37(7) (July, 2009), pp. 2763–2773; Benjamin K. Sovacool and Kelly Sovacool, "Preventing National Electricity–Water Crisis Areas in the United States," *Columbia Journal of Environmental Law* 34(2) (July, 2009), pp. 333–393; Benjamin K. Sovacool, *The Future of*

Nuclear Power: Boon or Blunder? (Washington, D.C.: Brookings Institution and Global Public Policy Initiative Policy Paper, March 2010); Benjamin K. Sovacool, "Critically Weighing the Costs and Benefits of a Nuclear Renaissance," *Journal of Integrative Environmental Science* 7(2) (June, 2010), pp. 105–122; Benjamin K. Sovacool and Anthony D'Agostino, "Nuclear Renaissance: A Flawed Proposition," *Chemical Engineering Progress* 106(7) (July, 2010), pp. 29–35; Benjamin K. Sovacool, "A Critical Evaluation of Nuclear Power and Renewable Energy in Asia," *Journal of Contemporary Asia* 40(3) (August, 2010), pp. 369–400; Benjamin K. Sovacool, "Exploring the Hypothetical Limits to a Nuclear and Renewable Electricity Future," *International Journal of Energy Research* 34 (November, 2010), pp. 1183–1194; and Ann Florini and Benjamin K. Sovacool, "Bridging the Gaps in Global Energy Governance," *Global Governance* 17(1) (in press, 2010).

Contents

1

Introduction: Exploring the "Faustian Bargain"

The economist John Maynard Keynes is reputed to have once said, "If a thing is not worth doing, it is not worth doing well." This book contends that nuclear power is one such thing not worth doing, well or otherwise. The basic premise behind a nuclear renaissance is wrong, whether one looks at it technically, economically, environmentally, or sociopolitically.

To make this case, the pages that follow conclusively show that new nuclear power plants are excessively capital-intensive, take years to build, are prone to cost overruns, and are economically competitive only when significantly subsidized. The history of operating performance shows an unacceptable rate of incidents and accidents that will grow in proportion with greater nuclear power generation. Secondary reserves of uranium will likely be exhausted before the end of the coming decade and high-quality reserves of primary uranium are already hard to find, contributing to rising and volatile fuel prices. Uranium mines are extremely hazardous to workers, and consistently discharge toxic tailings and radioactive dust that then become inhaled and ingested by nearby residents. The consequences of the nuclear fuel cycle to global water supply and land are disastrous, with a legacy of contaminating water tables and aquifers with tritium and other radioactive compounds. When quantified according to the best scientific studies to date, these social and environmental damages amount to more than US$220 billion every year. Nuclear facilities are attractive targets for terrorism, and they produce hazardous and radioactive material that can be used to make weapons. Accidents at existing nuclear power plants claimed more lives than all commercial airline accidents in the United States from 1982 to 2010, and epidemiological studies suggest that the global reactor fleet is responsible for 3,780 premature deaths and 1,253 cancers every year.

Even climate change — an area the nuclear industry has been quick to rally around — presents an issue that nuclear power plants are ill-suited to address. Within a few decades, the carbon footprint of nuclear plants will worsen to be equivalent to that of some fossil fuel-based sources of electricity. Electricity generation is only responsible for 40% of global greenhouse gas emissions. Nuclear plants must run steadily rather than with widely varying loads, as other power plants do. Nuclear units are too big for many small countries or rural areas. Moreover, nuclear power incurs higher costs than competing alternatives per unit of net carbon dioxide displaced, meaning that every dollar invested in nuclear expansion buys less carbon reduction than if that money was spent on other readily available solutions.[1] One study, for example, found that each dollar invested in energy efficiency displaces nearly seven times as much carbon dioxide as a dollar invested in nuclear power.[2]

Taken together, this book argues that these problems make a compelling case for avoiding a nuclear renaissance and instead investing in energy efficiency and renewable energy. Energy efficiency efforts and programs have been shown, time and time again, to be the cheapest, quickest, and environmentally friendliest approach to cutting emissions and improving energy security. Renewable power generators, in contrast to nuclear units, reduce dependence on foreign sources of uranium and decentralize electricity supply so that an accidental or intentional outage would have a more limited impact than the outage of larger nuclear facilities. Most significantly, renewable power technologies have environmental benefits because they generate electricity without relying on the extraction of uranium and its associated digging, drilling, mining, leaching, enrichment, transporting, and storage. As a result, energy efficiency and renewable energy technologies provide a much better alternative than nuclear power plants.

These collective technical, economic, environmental, and sociopolitical problems could be why the physicist Alvin M. Weinberg compared nuclear power to a "Faustian bargain." It creates an unbreakable commitment whereby society receives electricity only in exchange for yielding political power to a small cadre of technocrats and national security agencies as well as long-lasting nuclear waste. Plutonium, not generally found in nature and a byproduct of nuclear power generation, takes 240,000 years to become stable; whereas Stonehenge is slightly more than 4,000 years old, *Homo sapiens* migrated to Europe and Australia 40,000 years ago, and our species has been on earth for only 230,000 years.[3] Unlike Faust, who was ultimately able to renege on his deal with the devil, with nuclear energy society binds itself in perpetuity "to the remarkable belief that it can devise social institutions that are stable for periods equivalent to geologic ages."[4] It is far better to save energy or generate electricity by using wind, waves, sunlight, plants, or heat than by using nuclear power, which produces waste that will persist longer than our species has existed or that will last longer than our civilization has practiced Catholicism or cultivated agriculture.

What Makes This Book Different

Three things make this book unique: its critical focus, its interdisciplinary methodology, and its holistic emphasis on multiple dimensions of the nuclear fuel cycle.

First, this book will invariably offend advocates of nuclear power, for it is unabashedly one-sided and critical. The author is concerned about the dearth of recent literature critiquing nuclear power, with a host of books advancing the benefits of a nuclear renaissance without exploring its costs, to say nothing of the public relations material churned out by the industry and pro-nuclear associations. Libraries, webpages, and blogs are practically littered with academic and quasi-academic treatises espousing the promise of nuclear power, but downplay or ignore the risks and disadvantages.[5] This book is, first and foremost, an attempt to politicize the issue of nuclear power again, to provide some balance to the oversimplified and incomplete discourse currently proliferating on the Internet and through the mimetic commentary in the media. Yes, the book is therefore biased, but it is an informed bias based on years of careful study and one reached dispassionately; the author has no stake in the nuclear debate, nothing to personally gain or lose.

Second, to explore the challenges to a nuclear renaissance, the book draws from a rich array of interdisciplinary sources. The scope of literature surveyed includes not only the fields of public policy and economics but also technical reports and studies from the disciplines of nuclear engineering, ethics, law, science and technology, anthropology, sociology, and psychology. Special care has been taken to include as many peer-reviewed, academic, scientifically valid articles as possible, and not studies (pro or con) funded by either side of the nuclear debate.

Third, the book presents a multidimensional or comprehensive focus in myriad ways. The book investigates the entire nuclear fuel cycle, looking not only at nuclear reactors and spent fuel repositories but also at construction and decommissioning practices alongside uranium mines and mills as well as enrichment and fabrication facilities. It analyzes not only economic and technical issues but also environmental, social, political, and cultural ones, such as:

- Technical risks related to new designs, ongoing nuclear research, safety and accidents, as well as shortages of high-quality uranium fuel;
- Economic risks related to capital intensity, materials, labor, and long construction lead times;
- Environmental risks related to climate change, waste storage, uranium milling and mining, and water use; and
- Political risks related to weapons proliferation, accidents, and terrorism along with social risks related to employment, democracy, and marginalization.

Also, instead of focusing on nuclear energy in North America or Europe, it looks at the global status of the industry including South America and Asia in addition to the United States and the European Union.

Preview of Chapters

To make the case that a nuclear renaissance is not worth doing, the book proceeds as follows.

Chapter 2, "The Nuclear Industry: 'Smoking Cigarettes to Keep the Weight Off,'" explores the status of the industry, the basics of the nuclear fuel cycle, and plans for expansion. It briefly describes the various aspects of the nuclear fuel cycle and outlines current approaches to nuclear research and development, with a special emphasis on Generation IV nuclear reactors. It notes that the rapidly rising demand for electricity, dire warnings about climate change, and a desire to keep electricity prices low have motivated growth in the nuclear industry, and the chapter outlines the contours of a possible global expansion of nuclear power plants.

Chapter 3, "Safety and Reliability: Dealing with 'Normal Accidents,'" investigates many of the technical barriers to a nuclear renaissance. These include safety problems and accidents, shortages of materials and skilled labor, and a declining energy payback due to the lack of high-grade fuel.

Chapter 4, "Unfavorable Economics: 'Too Costly to Matter,'" describes seven categories of costs that make new nuclear power plants unusually expensive. Complex designs and long construction lead times contribute to high production and operating costs. The reprocessing of nuclear fuel adds significantly to the price tag of nuclear power, along with immense expenses related to waste storage, decommissioning, guarding of nuclear facilities, and new research and development.

Chapter 5, "Environmental Damages: 'Cutting Butter with a Chainsaw,'" discusses how uranium mining and milling, onsite waste storage facilities, and spent fuel repositories damage and degrade land. It also illustrates how the nuclear fuel cycle uses and contaminates water supplies, emits significant amounts of greenhouse gases, and places populations living near nuclear facilities at grave risk to medical and health problems.

Chapter 6, "Political and Social Concerns: 'Broken Plowshare,'" elaborates on a collection of security-related problems including transmission and distribution vulnerability, plant and reactor insecurity, weapons proliferation, military conflict, and maritime and land transport of fuel and fissile material. It also touches on how nuclear institutions interact negatively with democratic ideals and contribute to the marginalization of communities living near nuclear facilities.

Chapter 7, "Energy Efficiency and Renewable Energy: 'The Fire Extinguisher,'" describes two much better alternatives to nuclear energy: energy efficiency and renewable power sources such as wind, solar, hydroelectric, geothermal, and biomass. These smaller and more environmentally friendly practices and generators cost less to construct, produce power in smaller increments, and need not rely on continuous government subsidies. They generate little to no waste, have less greenhouse gas emissions per unit of electricity produced, and do not substantially contribute to the risk of accidents and weapons proliferation.

Chapter 8, "The 'Self-Limiting' Future of Nuclear Power," concludes the book by explaining why, if nuclear power has such immense costs, governments continue to support it. It offers explanations related to market failure and externalities, the socialization of risk, and hubris and technological fantasy to account for the attractiveness of nuclear power despite its enduring challenges.

The Stalled Nuclear Renaissance

What are we to make of claims of a nuclear renaissance then? The International Atomic Energy Agency (IAEA) reports that more than 60 countries have formally expressed an interest in introducing nuclear power to their energy sectors.[6] Even countries as diverse as Bangladesh, Belarus, Indonesia, Jordan, Myanmar, and Zimbabwe have either started or discussed starting nuclear energy programs, along with the better-known attempts in Iran, Iraq, and North Korea.[7]

Yet when all is said and done, the intricate challenges described in the chapters of this book make it likely that a true nuclear renaissance will never occur. Even if readers remain unconvinced about the arguments advanced against nuclear power, as well as those in favor of cleaner sources of energy, investors appear persuaded by them. As Chapter 7 forcefully argues, if one must build sources of electricity supply, nuclear reactors are a poor alternative compared to a host of renewable resources. Already today, the cheapest sources of electricity supply for markets in the US and the EU are wind turbines and landfill gas capture. Chapters 3 and 7 reveal that wind, biomass, geothermal, and hydroelectric power plants produce electricity for about 5–7 cents per kWh when subsidies are excluded. They, like energy efficiency, do so without the need for uranium, without risking catastrophic accidents, and without producing hazardous waste. Nuclear plants, when subsidies are excluded, produce power at up to 40 cents per kWh.

It is easy to do the math, but even if readers are not convinced, consider how investors the world over have been making their decisions. Renewable energy investment globally surpassed US$120 billion in 2008, a fourfold increase from 2004. Over the same period, solar energy grew by 600% and wind energy by 250%. Non-hydroelectric renewable resources as a whole grew at an annual rate of 23% in 2008; wind energy alone added 20 GW that year, yet Figure 1 shows that nuclear energy additions have stagnated at about 2 GW per year.[8] Table 1 also reveals that, from 2007 to 2008, both the US and the EU added more renewable capacity than conventional coal, oil, and nuclear capacity; and Table 2 shows that countries

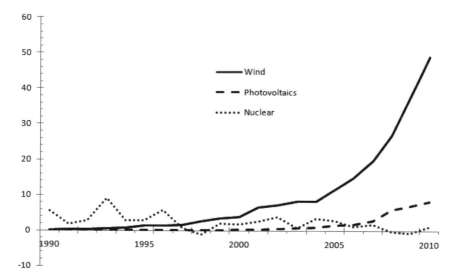

Figure 1: **Net Additions to Worldwide Electricity Generation for New Renewable and Nuclear Power Plants, 1990–2010 (in GW)[9]**

Table 1:   Global Investment in Renewable Sources of Electricity Supply, 2008

	Production/ Capacity (2008)	Annual Growth Rate (2007–2008)	Annual Investment (2008)
Wind	121 GW	29%	US$49 billion
Solar photovoltaics (PV)	13 GW	70%	US$39 billion
Concentrated solar power (CSP)	0.5 GW	6%	—
Small hydro	85 GW	8%	US$6 billion
Large hydro	860 GW	4%	US$42.5 billion
Geothermal power	10 GW	4%	US$2 billion
Biomass power	2 GW	4%	US$2 billion

Table 2: Top Five Countries for Renewable Energy Growth and Cumulative Investment, 2008

	#1	#2	#3	#4	#5
Annual Growth					
New capacity investment (all renewables)	United States	Spain	China	Germany	Brazil
Wind	United States	China	India	Germany	Spain
Solar PV	Spain	Germany	United States, South Korea, Japan, Italy		
Solar hot water/heat	China	Turkey	Germany	Brazil	France
Ethanol production	United States	Brazil	China	France	Canada
Biodiesel production	Germany	United States	France	Argentina	Brazil
Total Capacity					
All renewables	China	United States	Germany	Spain	India
Small hydro	China	Japan	United States	Italy	Brazil
Wind	United States	Germany	Spain	China	India
Biomass power	United States	Brazil	Philippines	Germany, Finland, Sweden	
Geothermal power	United States	Philippines	Indonesia	Mexico	Italy
Solar PV	Germany	Spain	Japan	United States	South Korea
Solar hot water/heat	China	Turkey	Germany	Japan	Israel

Note: Solar includes only grid-connected solar PV.

in almost every continent significantly invested in renewables. In contrast, the global share of nuclear energy did not grow at all; it actually *shrank* by about 1%, and the amount invested in 2008 was one-tenth the amount invested in renewables and energy efficiency. During the decade 1996–2006, while total global primary energy consumption increased by 26%, the aggregate increase of installed nuclear power was a mere 15%.

All the while, the number of operating reactors is declining (from 438 in 2002 to only 432 in 2010) and the average age of the nuclear fleet is growing. Figure 2 shows that the amount of nuclear electricity generated worldwide decreased by 2% from 2007 to 2008, and in the EU it dropped by 6% — more than in any other year since the first reactor was connected to the Soviet grid in 1954.[10] Figure 3 shows that most nuclear plants were installed in 1984 and 1985, and that the average age of a nuclear facility is now 25 years.[11] IAEA and World Nuclear Association statistics showing a large number of reactors "in operation" or "in construction" are also misleading. Seventeen of those reactors in Germany, Japan, India, and the United Kingdom did not generate any power at all in 2009 (yet were still counted as installed capacity).[12] Fifty-two reactors are currently "under construction," but several are orders from previous eras that have become bogged down in delays. Some are partially mothballed reactors that are not

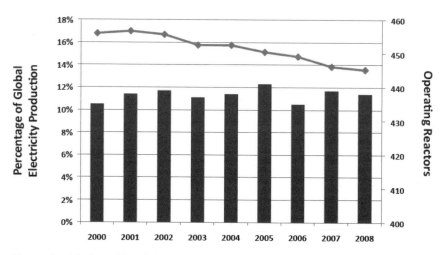

Figure 2:   Nuclear Electricity Generation and Number of Worldwide Reactors, 2000–2008

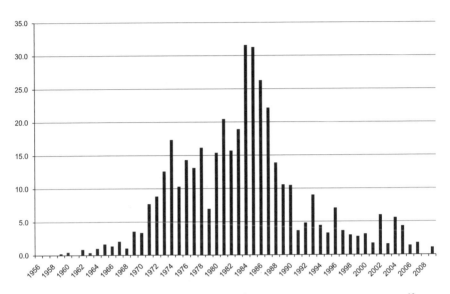

Figure 3:   Average Age of Global Reactor Fleet (Year reactors were installed)[13]

likely to be completed again, such as the 692-MW plant in Argentina. Others are only small and experimental research reactors, like those in Russia. Thirty are taking place in only four countries (China, India, Russia, and South Korea), while only one country (Iran) is building its first reactor. The reactors nearing completion are predominantly older-generation models (Generation II and II+). Lastly, a large number of reactors in construction do not even have an official start date, raising questions about whether construction will ever commence.

It is no surprise, then, that the Massachusetts Institute of Technology (MIT) conducted a study concluding that only by imposing a carbon tax of up to US$200 per ton on conventional power plants could advanced nuclear reactors be cost-competitive with existing conventional technologies.[14] Similarly, after assessing the recent spikes in construction cost, operational safety, radioactive waste disposal, and public acceptance, even economists at the pro-nuclear IAEA concluded that a global expansion of nuclear power is unlikely, and they projected that the global nuclear industry will likely lose market share by 11.6% by 2020.[15]

In short, investors and planners in Europe and the US are choosing renewable energy over new nuclear power plants. The private capital market is not investing in nuclear power; and without financing, the only purchases of new nuclear equipment are being made in Asia by central planners with a draw on the public purse. Amory Lovins has mused that the lesson seems to be that, in today's market, governments can have only about as many nuclear plants as they can force taxpayers to purchase.[16]

The barriers to a nuclear renaissance appear especially acute for developing countries, which face a unique set of challenges. The ability for countries such as Myanmar and Zimbabwe to operate nuclear facilities safely depends on their capacity to manage a large and complex technological system over many decades. Many developing countries — especially the least developed or lower-middle income countries — have a hard enough time struggling with corruption, the threat of terrorism, and civil unrest, and lack the capacity to meaningfully pursue new nuclear power plants. The physical infrastructure needed to support a plant is quite expansive and expensive, as noted in Chapters 2 and 3. Furthermore, the IAEA requires hundreds of targets that must be met for international accreditation; these include roads, a safe and secure reactor site, a large supply of water, and a robust transmission grid, which many developing countries lack. Accomplishing these supplemental but important criteria will likely cost billions of dollars above and beyond the capital cost of a plant. Finally, in the current financial climate, it is difficult for these countries to find financiers to cover the US$10–20 billion needed for a new reactor. Such countries often have a low per capita GDP, poor credit ratings, and trouble obtaining commercial loans.[17]

Two examples from Southeast Asia are telling: Indonesia and Vietnam. In Indonesia, hardly a least developed country, all three of its research reactors have experienced serious technical problems due to lack of maintenance and inadequate operational procedures. The TRIGA Mark II reactor showed bubbling in its core after an upgrade in 2002 when a fuel element jammed. The Kartini-P3TM reactor was found to have a corroded reactor liner in three areas, prompting the IAEA to intervene to decommission and clean up the plant. The Siwabessy Multipurpose Research Reactor has seen the corrosion of its cooling system and a crack in its reactor liner.[18]

Vietnam is also emblematic of the logistical challenges facing a developing country trying to build its first nuclear plant. The country lacks qualified nuclear engineers and is rushing to train 500 by 2020. The sites selected for the first two nuclear power plants face the sea and are therefore at risk of rising sea levels associated with climate change. Vietnam has no repository in place for the storage of nuclear waste. Members of Parliament have expressed concern that the plants will be too expensive and will not be an efficient use of revenue. Arms control experts have also warned that a nuclear program could enable Vietnam to pursue nuclear weapons.[19]

Clearly, for developing countries such as Indonesia and Vietnam, maintaining research reactors and building commercial nuclear power plants remain an incredible technical challenge. As the chapters to come show, similar obstacles face all countries and communities trying to push nuclear power — barriers which cast serious doubt over any type of nuclear renaissance. Arthur Conan Doyle's Sherlock Holmes once said: "It is a capital mistake to theorize before one has data. Insensibly one begins to twist facts to suit theories, instead of theories to suit facts."[20] Advocates of nuclear energy must consider all of the data before they make a commitment to new reactors. Ignoring serious risks relating to burgeoning operating costs, greenhouse gases, water use, childhood cancer, insecure fuel supplies, and nuclear waste storage will not make them go away.

Endnotes

[1] Amory B. Lovins, "Nuclear Power: Economics and Climate Protection Potential" (Snowmass, CO: Rocky Mountain Institute, 2005, E05-08). Specifically, Lovins calculates that every 10 cents spent to buy a single kWh of nuclear electricity, assuming subsidies and regulation in the US, could have purchased 1.2–1.7 kWh of wind power, 0.9–1.7 kWh of gas, 2.2–6.5 kWh of building-scale cogeneration, or 10 kWh or more of energy efficiency. Put another way, nuclear power saves as little as half as much carbon per dollar as wind power or cogeneration, and from several-fold to at least tenfold less carbon per dollar than end-use energy efficiency.

[2] Bill Keepin and Gregory Kats, "Greenhouse Warming: Comparative Analysis of Nuclear and Efficiency Abatement Strategies," *Energy Policy* 16 (1988), p. 552.

[3] Greenpeace, *Nuclear Power: A Dangerous Waste of Time* (Amsterdam: Greenpeace International, 2009).

[4] Alvin M. Weinberg, "Social Institutions and Nuclear Energy," in *Energy and the Way We Live* (1980), pp. 311–312.

[5] For a small sample of these works, readers can visit their local library or the Internet and search for books on "nuclear power" or "atomic energy." Roughly four out of every five tend to be uncritically positive. See, for example, Robert C. Morris, *The Environmental Case for Nuclear Power: Economic, Medical, and Political Considerations* (New York: Continuum, 2000); Ian Hore-Lacy, *Nuclear Energy in the 21st Century* (New York: Academic Press, 2006); Alan M. Herbst and George W. Hopley, *Nuclear Energy Now: Why the Time Has Come for the World's Most Misunderstood Energy Source* (New York: Wiley, 2007); Gwyneth Cravens and Richard Rhodes, *Power to Save the World: The Truth About Nuclear Energy* (New York: Vintage, 2008); William Tucker, *Terrestrial Energy: How Nuclear Energy Will Lead the Green Revolution and End America's Energy Odyssey* (New York: Bartleby Press, 2008); and James Mahaffey, *Atomic Awakening: A New Look at the History and Future of Nuclear Power* (New York: Pegasus, 2009).

[6] Interview with David Waller, Deputy Director General of the IAEA, Vienna, Austria, May 6, 2009.

[7] World Nuclear Association, "World Nuclear Power Reactors 2007–09 and Uranium Requirements," May 1, 2009.

[8] Antony Froggatt and Mycle Schneider, *Systems for Change: Nuclear Power vs Energy Efficiency + Renewables* (London: Heinrich-Böll-Stiftung, March 2010).

[9] *Ibid.*

[10] Mycle Schneider, "2008 World Nuclear Industry Status Report: Global Nuclear Power," *Bulletin of Atomic Scientists*, September 16, 2008.

[11] Trevor Findlay, *The Future of Nuclear Energy to 2030 and Its Implications for Safety, Security, and Nonproliferation* (Waterloo, Ontario: Centre for International Governance Innovation, 2010).

[12] Mycle Schneider, Steve Thomas, Antony Froggatt, and Doug Koplow, *The World Nuclear Industry Status Report 2009* (Paris: German Federal Ministry of Environment, Nature Conservation and Reactor Safety, August 2009, UM0901290).

[13] Source: Antony Froggatt, "Nuclear Self-Sufficiency — Can Nuclear Power Pave the Road Towards Energy Independence?," Presentation to the "Towards a Nuclear Power Renaissance" Conference in Potsdam, Germany, March 4–5, 2010.

[14] Massachusetts Institute of Technology, *The Future of Nuclear Power: An Interdisciplinary MIT Study* (Cambridge, MA: MIT, 2003).

[15] Ferenc L. Toth and Hans-Holger Rogner, "Oil and Nuclear Power: Past, Present, and Future," *Energy Economics* 28 (2006), p. 21.

[16] Lovins (2005).

[17] Findlay (2010).

[18] International Institute for Strategic Studies, *Preventing Nuclear Dangers in Southeast Asia and Australasia* (London: IISS, September 2009), pp. 20–21.

[19] Ta Minh Tuan, "Implications of Vietnam's Planned Nuclear Power Station," *Energy Studies Institute Bulletin on Energy Trends and Development* 2(3) (December, 2009), pp. 5–6.

[20] Arthur Conan Doyle, *A Scandal in Bohemia* (1891), p. 78.

2

The Nuclear Industry: "Smoking Cigarettes to Keep the Weight Off"

The nuclear era began with a whimper, not a bang, on December 7, 1942. Amidst the polished wooden floors of a war-appropriated squash court at the University of Chicago, Enrico Fermi inserted about 50 tons of uranium oxide into 400 carefully constructed graphite blocks. A small puff of heat exhibited the first self-sustaining nuclear reaction, many bottles of Chianti were consumed, and nuclear energy was born.[1]

The relative simplicity of the first human-induced nuclear reaction, however, obscures just how complicated fission is at producing electricity. After exploring the current status of the nuclear industry, this chapter describes the basic components of the nuclear fuel cycle: uranium mining, milling, and conversion and enrichment; reactor design and construction; operation; fuel processing, storage, and waste sequestration; and decommissioning. It then elaborates some of the drivers behind the current push for the expansion of nuclear power.

Current Status of the Nuclear Industry

Ever since the first experimental nuclear reactor produced electricity in 1951 in Idaho, the first commercial nuclear facility went online in 1956 at Calder Hall in the United Kingdom, the first demonstration plant in the

15

Figure 1:   The 60-MW Nuclear Power Plant at Shippingport, Pennsylvania

United States was completed at Shippingport in 1957 (Figure 1), and the first American commercial nuclear plant was built in 1963, nuclear energy has been touted as the modern solution to the world's growing demand for energy.

As Table 1 shows, as of April 2010, a total of 31 countries operated 432 nuclear reactors constituting 369 GW of installed capacity. Together, these reactors produced about 15% of the world's electricity, or 2,560.6 billion kWh. In the US alone, which had almost one-quarter of the world's reactors, nuclear facilities accounted for almost 20% of national electricity generation. In France, 76% of electricity came from nuclear sources; and nuclear energy contributed more than 20% of national power production in Germany, Japan, South Korea, Sweden, and Ukraine. As Figure 2 shows, the US, France, Japan, Russia, South Korea, and Germany produced almost three quarters of the nuclear electricity in the world in 2008. Collectively, the global fleet of commercial nuclear power plants needed about 68,000 metric tons of uranium to operate for one year, and it represented roughly 12,600 reactor-years of experience. Moreover, 56 countries operated 284 research reactors and a further 220

Table 1: Commercial Nuclear Power Generation, Reactors, and Fuel Requirements[2]

Country	Nuclear Electricity Generation (2008)		Operating Reactors (April 2010)		Reactors Under Construction (April 2010)		Uranium Required (2010) (metric tons)	Technology	Supplier(s)
	billion kWh	percent electricity	No.	MWe	No.	MWe			
Argentina	6.8	6.2	2	935	1	692	123	Heavy water reactor	Siemens, Atomic Energy of Canada Limited
Armenia	2.3	39.4	1	376	0	0	55	Vodo-Vodyanoi Energetichesky Reactor	Russia
Belgium	43.4	53.8	7	5,943	0	0	1,052	Pressurized water reactor	Framatome
Brazil	14.0	3.1	2	1,901	0	0	311	Pressurized water reactor	Westinghouse, Siemens
Bulgaria	14.7	32.9	2	1,906	0	0	272	Vodo-Vodyanoi Energetichesky Reactor	Russia
Canada	88.6	14.8	18	12,679	2	1,500	1,675	Heavy water reactor	Atomic Energy of Canada Limited
China	65.3	2.2	11	8,587	21	22,960	2,875	Pressurized water reactor, heavy water reactor,	Framatome, Atomic Energy of Canada Limited, China,

(Continued)

Table 1: *(Continued)*

Country	Nuclear Electricity Generation (2008) billion kWh	percent electricity	Operating Reactors (April 2010) No.	MWe	Reactors Under Construction (April 2010) No.	MWe	Uranium Required (2010) (metric tons)	Technology	Supplier(s)
								Vodo-Vodyanoi Energetichesky Reactor	Russia
Czech Republic	25.0	32.5	6	3,686	0	0	678	Vodo-Vodyanoi Energetichesky Reactor	Russia
Finland	22.0	29.7	4	2,696	1	1,600	1,149	Boiling water reactor, pressurized water reactor, Vodo-Vodyanoi Energetichesky Reactor	Russia, Asea, Westinghouse
France	418.3	76.2	58	63,236	1	1,630	10,153	Pressurized water reactor	Framatome
Germany	140.9	28.3	17	20,339	0	0	3,453	Pressurized water reactor, boiling water reactor	Siemens

(Continued)

Table 1: (Continued)

Country	Nuclear Electricity Generation (2008)		Operating Reactors (April 2010)		Reactors Under Construction (April 2010)		Uranium Required (2010)	Technology	Supplier(s)
	billion kWh	percent electricity	No.	MWe	No.	MWe	(metric tons)		
Hungary	14.0	37.2	4	1,880	0	0	295	Vodo-Vodyanoi Energetichesky Reactor	Russia
India	13.2	2.0	19	4,183	4	2,572	908	Heavy water reactor, Vodo-Vodyanoi Energetichesky Reactor, fast breeder reactor	Atomic Energy of Canada Limited, India, Russia
Iran	0	0	0	0	1	915	148	Vodo-Vodyanoi Energetichesky Reactor	Russia
Japan	240.5	24.9	54	47,102	1	1,373	8,003	Boiling water reactor, pressurized water reactor	Westinghouse, General Electric, Hitachi, Mitsubishi, Toshiba

(Continued)

Table 1: (*Continued*)

Country	Nuclear Electricity Generation (2008) billion kWh	percent electricity	Operating Reactors (April 2010) No.	MWe	Reactors Under Construction (April 2010) No.	MWe	Uranium Required (2010) (metric tons)	Technology	Supplier(s)
Lithuania	9.1	72.9	0	0	0	0	0	Reactor Bolshoy Moshchnosty Kanalny	Russia
Mexico	9.4	4.0	2	1,310	0	0	253	Boiling water reactor	General Electric
Netherlands	3.9	3.8	1	485	0	0	107	Pressurized water reactor	Siemens
Pakistan	1.7	1.9	2	400	1	300	68	Heavy water reactor, pressurized water reactor	Canada, China
Romania	7.1	17.5	2	1,310	0	0	175	Heavy water reactor	Atomic Energy of Canada Limited
Russia	152.1	16.9	32	22,811	8	6,380	4,135	Reactor Bolshoy Moshchnosty Kanalny, Vodo-Vodyanoi Energetichesky Reactor	Russia

(*Continued*)

Table 1: (*Continued*)

Country	Nuclear Electricity Generation (2008)		Operating Reactors (April 2010)		Reactors Under Construction (April 2010)		Uranium Required (2010) (metric tons)	Technology	Supplier(s)
	billion kWh	percent electricity	No.	MWe	No.	MWe			
Slovakia	15.5	56.4	4	1,760	2	840	269	Vodo-Vodyanoi Energetichesky Reactor	Russia
Slovenia	6.0	41.7	1	696	0	0	145	Pressurized water reactor	Westinghouse
South Africa	12.7	5.3	2	1,842	0	0	321	Pressurized water reactor	Framatome
South Korea	144.3	35.6	20	17,716	6	6,700	3,804	Pressurized water reactor, heavy water reactor	Atomic Energy of Canada Limited, Westinghouse, South Korea
Spain	56.4	18.3	8	7,448	0	0	1,458	Pressurized water reactor, boiling water reactor	Westinghouse, General Electric, Siemens
Sweden	61.3	42.0	10	9,399	0	0	1,537	Pressurized water reactor, boiling water reactor	Westinghouse, Asea

(*Continued*)

Table 1: (*Continued*)

Country	Nuclear Electricity Generation (2008)		Operating Reactors (April 2010)		Reactors Under Construction (April 2010)		Uranium Required (2010)	Technology	Supplier(s)
	billion kWh	percent electricity	No.	MWe	No.	MWe	(metric tons)		
Switzerland	26.3	39.2	5	3,252	0	0	557	Pressurized water reactor, boiling water reactor	Westinghouse, General Electric, Siemens
Ukraine	84.3	47.4	15	13,168	0	0	2,031	Vodo-Vodyanoi Energetichesky Reactor	Russia
United Kingdom	52.5	13.5	19	11,035	0	0	2,235	Gas-cooled reactor, pressurized water reactor	United Kingdom, Westinghouse
United States	809.0	19.7	104	101,119	1	1,180	19,538	Pressurized water reactor, boiling water reactor	Westinghouse, General Electric, Babcock & Wilcox, Combustion Engineering
Total	2,560.6	—	432	369,200	50	48,642	67,783		

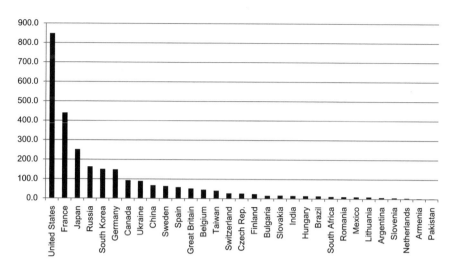

Figure 2:   Global Nuclear Electricity Production (TWh) in 2008[3]

reactors were used to power ships and submarines, bringing the world total to almost 1,000 reactors.

As Table 1 also reveals, commercial operators rely on a mix of technologies and suppliers — mostly light water reactors (especially pressurized water reactors) and heavy water reactors — to produce electricity. Light water reactors constitute more than 80% of the world's fleet. Light water reactors in this context use ordinary water, so-called "light" water in the early days of the industry to distinguish it from "heavy" water (which replaces a hydrogen atom with deuterium). Light water reactors have two basic types: a pressurized water reactor and a boiling water reactor. A pressurized water reactor has a nuclear core that heats water under pressure and passes it through a boiler; in a boiling water reactor, the water boils above the nuclear reaction and directly creates steam to turn a turbine.[4]

Nuclear engineers often describe four generations of nuclear plant design. The first generation refers to the experimental reactors designed in the 1940s and 1950s. These rather small "Atoms for Peace"-era plants are now almost all shut down. Only six Generation I units were still operating in 2007 — a series of small, 250-MW, gas-cooled nuclear power plants in the UK.

The second generation of nuclear plants refers to most commercial reactors now in operation, including the light water reactor fleet found in the US and Europe, predominately comprised of pressurized water reactors and boiling water reactors. These reactors were mostly designed in the 1960s and built in the 1970s.

The third generation encompasses advanced reactor designs that operate at slightly higher temperatures or according to different designs, such as pebble bed modular reactors, Canadian deuterium uranium reactors, European pressurized water reactors, and advanced boiling water reactors. These advanced reactors, sometimes referred to as "Generation III" or "Generation III+" technology, emerged from public-private research in the 1980s and early 1990s. While Generation III reactors are not currently used widely by the industry, their use is expected to grow significantly between 2020 and 2040. The difference between Generation III and Generation IV designs is sometimes blurred by Generation III proponents attempting to receive Generation IV research funds.

Research on the fourth generation of nuclear reactors, often called "Generation IV" systems, began in the late 1990s under the Advanced Fuel Cycle Initiative, previously called the Advanced Accelerator Applications Program. Under the Advanced Fuel Cycle Initiative, the US started researching advanced reactor designs and fuel cycles along with Belgium, China, the Czech Republic, France, Germany, Hungary, India, Japan, the Netherlands, Poland, the Russian Federation, South Korea, Switzerland, and the European Commission. The program morphed into the Nuclear Energy Research Initiative (NERI) in 1999, a project headed by the US Department of Energy (DOE) to explore research in four areas (explained in Table 2). Under the NERI, the DOE alone sponsored 46 research projects involving national laboratories, universities, and industry and foreign research partners.

The DOE's approach to Generation IV research transformed once again in 2002, when President George W. Bush announced a nuclear program aptly called the "Generation IV International Forum" (GIF). At the heart of the GIF lies the Global Nuclear Energy Partnership, a program created in 2006 to promote nuclear energy abroad by exploring export opportunities for American technology firms. Ten countries

Table 2:   Objectives of "Next Generation" Nuclear Research and Development

Area	Objective
Construction	Moving away from onsite construction of power plants to a more standardized manufacturing approach with simplified designs that would be more suited to mass production
Proliferation resistance	Creating fuel core designs that operate for at least 15 years without refueling to minimize the risk of theft of fissile materials
Safety	Improving operational procedures and maintenance requirements to minimize the incidence of human operator error
Waste disposal	Designing new fuel cycles to minimize the creation of nuclear waste and operate on alternative forms of fuel

currently announce and share their research efforts annually at the GIF: Argentina, Brazil, Canada, France, Japan, South Africa, South Korea, Switzerland, the UK, and the US. Generation IV research initially started by considering a slate of 20 different reactor designs, but has since been narrowed down to only six designs: very-high-temperature reactors, gas-cooled fast reactors, sodium-cooled fast reactors, supercritical water-cooled reactors, lead-cooled reactors, and molten salt reactors. These designs, while differing in specific engineering, have six common themes: (1) they are intended to produce reactors operating either at very high temperatures or in a fast neutron/breeder fuel cycle that attempts to recycle spent fuel; (2) they attempt to improve the environmental performance of reactors by minimizing the need for mined uranium and lessening the environmental footprint of power plants; (3) they plan to improve waste management by recycling or minimizing the fuel that they do use; (4) they try to enhance proliferation resistance by making it impossible to steal weapons-grade material; (5) they intend to improve safety and reliability; and (6) they attempt to minimize financial risk and improve the economics of plant construction and operation.

In short, the theory is that future Generation IV nuclear technology would operate differently than conventional units by utilizing fuel cycles with higher temperatures or different forms of fuel, minimizing environmental damage and the creation of waste, decreasing the amount of fissile material from the fuel cycle available for weapons, improving safety, and lowering the price of nuclear power plants. However, Generation IV reactors are also the furthest from commercialization. They are completely experimental, with engineers and scientists still working out theoretical concepts, most of which have not been proven in practice. The next stage in Generation IV research, if possible, would likely be the construction of experimental reactors around 2015 or 2020. Then, if successful, commercialization and wide-scale deployment of Generation IV technologies would begin by 2040 at the earliest. (As later chapters note, though, Generation IV reactors have been called "paper reactors" because it is likely they may only exist on paper.)

The Basics of the Nuclear Fuel Cycle

Engineers generally classify nuclear fuel cycles into two types: once-through and closed. Conventional reactors operating on a once-through mode discharge spent fuel directly into disposal. Reactors with reprocessing in a closed fuel cycle separate waste products from unused fissionable material so that it can be recycled as fuel. Reactors operating on closed cycles extend fuel supplies and have advantages in terms of storage of waste disposal, but have disadvantages in terms of cost, short-term reprocessing issues, proliferation risk, and fuel cycle safety.[5]

Despite these differences, both once-through and closed nuclear fuel cycles involve at least five interconnected stages: the front end of the cycle where uranium fuel is mined, milled, converted, enriched, and fabricated; the construction of the plant itself; the operation and maintenance of the facility; the back end of the cycle where spent fuel is conditioned, (re)processed, and stored; and a final stage where plants are decommissioned and abandoned mines are returned to their original state. All sorts of diagrams and figures have been drawn to illustrate the nuclear fuel cycle; two of my favorite are presented in Figures 3 and 4. As both figures indicate, the nuclear fuel cycle is long and complex, though not all stages of the fuel cycle are present in all reactors. Canadian-designed heavy water CANDU reactors, for instance, do not need uranium enrichment, as they rely on natural uranium as fuel.[6]

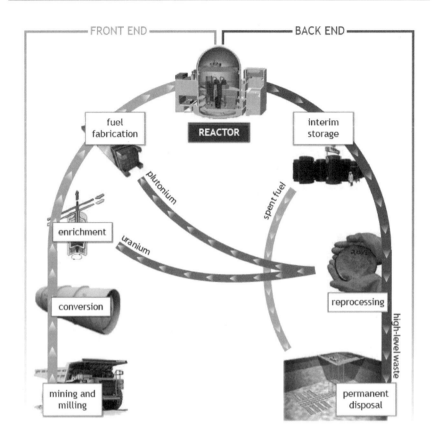

Figure 3:   Front-End/Back-End Representation of the Nuclear Fuel Cycle[7]

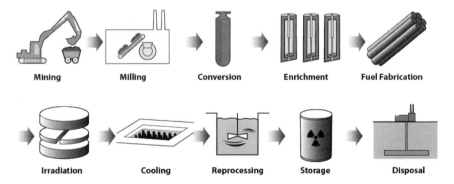

Figure 4:   Linear Representation of the Nuclear Fuel Cycle[8]

The Front End

The front end of the cycle begins with uranium, the primary fuel for nuclear power plants. It is widely distributed in the earth's crust and the ocean in minute quantities, with the exception of concentrations rich enough to constitute ore. Uranium is mined both at the surface and underground; and after it is extracted, it is crushed, ground into fine slurry, and leached in sulfuric acid. Uranium is then recovered from solution and concentrated into solid uranium oxide, often called "yellowcake," before it is converted into uranium hexafluoride and heated. Then, hexafluoride vapor is loaded into cylinders where it is cooled and condensed into a solid before undergoing enrichment through gaseous diffusion or gas centrifuge.

Uranium Mining

Starting at the mine, rich ores may embody concentrations of uranium oxide as high as 10%, but 0.2% or less is usual and most uranium producers will consider mining ores with concentrations higher than 0.0004%. A majority of the usable soft ore found in sandstone has a concentration between 0.2% and 0.01%; while hard ore found in granite has a lower uranium content, usually about 0.02% or less. Uranium mines are typically open-cast pits (up to 250 m deep) or underground. A third extraction technique involves subjecting natural uranium to *in situ* leaching, whereby hundreds of tons of sulfuric acid, nitric acid, and ammonia are injected into the strata and then pumped up again after 3–25 years, yielding uranium from treated rocks.

Uranium Milling

Mined uranium must undergo a series of metallurgical processes to crush, screen, and wash the ore, letting the heavy uranium settle as the lighter debris is funneled away. The next step is the mill, often situated near the mine, where acid or alkali baths leach the uranium out of the processed ore, producing a bright yellow powder called yellowcake, which is about 75% uranium oxide (its chemical form is U_3O_8). In cases where ores have a concentration of 0.1%, the milling must grind

1,000 tons of rock to extract one ton of yellowcake; both the oxide and the tailings (the 999 tons of remaining rock) remain radioactive, requiring treatment. Acids must be neutralized with limestone and made insoluble with phosphates, the environmental consequences of which are discussed in Chapter 5.[9]

Uranium Conversion and Enrichment

Next comes conversion and enrichment, where a series of chemical processes are conducted to remove the remaining impurities and to convert yellowcake into the necessary compounds for fuel pellets and assemblies. Enrichment is needed to increase the percentage of the isotope uranium-235 to higher levels needed for most types of reactors. Natural uranium contains about 0.7% uranium-235; the rest is mainly uranium-234 or uranium-238. In order to bring the concentration of uranium-235 up to at least 3.5% for typical commercial light water reactors or about 4–5% for other modern reactors, the oxide must be enriched. The process begins by converting uranium to uranium hexafluoride (UF_6 or "hex").

A variety of enrichment processes have been demonstrated in the laboratory, including:

- gaseous diffusion;
- thermal diffusion;
- gas centrifuge separation;
- atomic vapor laser isotope separation;
- molecular laser isotope separation;
- separation of isotopes by laser excitation;
- aerodynamic isotope separation;
- electromagnetic isotope separation;
- plasma separation; and
- chemical separation.

The two dominant commercial enrichment methods are gaseous diffusion and centrifuge separation.[10] Table 3 shows that 21 large uranium enrichment facilities were either operational or under construction as of late 2009.

Table 3:   **Large Uranium Enrichment Facilities**

Country	Name/Location	Process	Capacity (Thousands of Separative Work Units/Year)
Brazil	Resende Enrichment	Centrifuge	120
China	Lanzhou 2	Centrifuge	500
China	Shaanxi Enrichment Plant	Centrifuge	500
France	Georges Besse	Diffusion	10,800
France	Georges Besse II	Centrifuge	7,500
Germany	Gronau	Centrifuge	2,000
India	Rattehallib	Centrifuge	10
Iran	Natanz	Centrifuge	150
Japan	Rokkasho Enrichment Plant	Centrifuge	1,050
Netherlands	Almelo	Centrifuge	4,000
Pakistan	Kahutab	Centrifuge	20
Russia	Angarsk	Centrifuge	2,600
Russia	Nouvouralsk	Centrifuge	9,800
Russia	Zelenogorsk	Centrifuge	5,800
Russia	Seversk	Centrifuge	4,000
United Kingdom	Capenhurst	Centrifuge	5,000
United States	Paducah	Diffusion	11,300
United States	Portsmouth	Diffusion	7,400
United States	National Enrichment Facility	Centrifuge	3,000
United States	American Centrifuge Plant	Centrifuge	3,500
United States	Idaho Falls Enrichment Plant	Centrifuge	3,000

Gaseous diffusion, developed during the Second World War as part of the Manhattan Project, accounts for about 45% of world enrichment capacity. The diffusion process funnels hex through a series of porous membranes or diaphragms. The lighter uranium-235 molecules move faster than the uranium-238 molecules and have a slightly better chance of passing through the pores in the membrane. The process is repeated many times in a series of diffusion stages called a cascade, with the enriched UF_6 withdrawn from one end of the cascade and the depleted UF_6 removed at the other end. The gas must be processed through some 1,400 stages before it is properly enriched.

The gas centrifuge process, first demonstrated in the 1940s, feeds hex into a series of vacuum tubes, and accounts for about 45% of world enrichment capacity. When the rotors are spun rapidly, the heavier molecules with uranium-238 increase in concentration towards the outer edge of the cylinders, with a corresponding increase in uranium-235 concentration near the center. To separate the two isotopes, centrifuges rotate at very high speeds, with spinning cylinders moving at roughly one million times the acceleration of gravity.

In the US, the gaseous diffusion plant at Paducah, Kentucky, primarily does enrichment; while Europe and Russia utilize mostly centrifuge methods.[11] The remaining nuclear fuel (about 10%) comes from the recycling of nuclear weapons. After enrichment, about 85% of the oxide comes out as waste in the form of depleted hex, also known as enrichment tails, which must be stored. Each year, for instance, France creates 16,000 tons of enrichment tails that are then exported to Russia or added to the existing 200,000 tons of depleted uranium within the country.[12]

The 15% that emerges as enriched uranium is converted into ceramic pellets of uranium dioxide (UO_2), packed in zirconium alloy tubes, and bundled together to form fuel rod assemblies for reactors. Fuel fabrication involves tightly pressing UO_2 powder into small pellets and then sintering them to form a ceramic. Fuel elements and assemblies are often exposed to even more ionizing radiation through accelerators and cyclotrons — a process known as irradiation — to bombard atoms with particles so as to produce a chain reaction. To supply enough enriched fuel for a standard 1,000-MW reactor for 1 year, about 200 tons of natural uranium has to be processed.

Construction

The construction phase of the nuclear lifecycle involves the fabrication, transportation, and use of materials to build generators, turbines, cooling towers, control rooms, transformers, substations, and other infrastructure. A typical nuclear plant contains some 50 miles of piping (welded 25,000 times) and 900 miles of electrical cables. Thousands of electric motors, conduits, batteries, relays, switches, operating boards, condensers, and fuses are needed for the system to operate. Cooling systems necessitate valves, seals, drains, vents, gauges, and fittings. Structural supports, firewalls, radiation shields, spent fuel storage facilities, and emergency backup generators must remain in excellent condition. Temperatures, pressures, power levels, radiation levels, flow rates, cooling water chemistry, and equipment performance must all be constantly monitored. A common nuclear plant needs 170,000 tons of concrete, 32,000 tons of steel, 1,363 tons of copper, and a total of 205,464 tons of other materials. Many of these are energy- and carbon-intensive: 1 ton of aluminum has a carbon equivalent of more than 10,000 tons of CO_2; 1 ton of lithium, 44,000 tons; and 1 ton of silver, 913,000 tons.[13] Some of the constraints surrounding the construction of new nuclear power plants are discussed in Chapters 3 and 5.

Operation

The operation phase of the lifecycle encompasses the energy needed to manage the cooling and fuel cycles of the plant, as well as the energy needed for its maintenance and backup generators. Indirect energy use includes the provision of power during reactor outages, repairs, and shutdowns.

The heart of the operating nuclear facility is the reactor, which generates electricity through the fission, or splitting, of uranium and plutonium isotopes. The nuclear reaction inside a nuclear power plant is identical to a nuclear weapon explosion, except the reactor releases the energy slowly over months instead of in one awesome, terrifying moment. In a nuclear reactor, the fission process does not take place one atom at a time. Uranium has the rare and productive property that, when it is struck by a neutron, it splits into two and produces more neutrons. If one uranium-235 atom collides with an atom of uranium-238 (one of the other isotopes of uranium), it may stay there and induce a couple of decay cycles to produce plutonium-239. Plutonium-239, sharing the same property as uranium-235, splits when struck by neutrons to act as additional fuel. The process can be controlled by a moderator consisting of water or graphite to speed the reaction up, and neutron-absorbing control rods to slow it down. Most nuclear reactors around the world have a present lifetime of 30–40 years, but produce electricity at full power for no more than 24 years.[14]

Daily work at a typical nuclear power plant is carried out by between 200 and 1,000 people, although the number jumps significantly during refueling and maintenance. During some phases, planned work activities are truly immense, ranging from several hundred to several thousand operations and tasks per hour; and during an outage, more than 90,000 work orders can occur for one or more tasks.[15]

The Back End

The back end of the nuclear fuel cycle involves fuel processing, interim storage, and permanent sequestration of waste.

Fuel Processing and Reprocessing

Spent fuel must be conditioned for reactors operating on a once-through fuel cycle, and reprocessed for those employing a closed fuel cycle. Eventually, radioactive impurities such as barium and krypton, along with transuranic elements such as americium and neptunium, clog the uranium inducing a nuclear reaction. After a few years, fuel elements must be removed and fresh fuel rods inserted. In France and the UK, where reprocessing continues, spent uranium is stored for hopeful use at a later date in fast breeder reactors, plutonium is recycled into mixed oxide (MOX), and the remaining fissile waste is vitrified (i.e. chemically transformed into a glass to make the waste inert). This method of reprocessing — plutonium-uranium extraction (PUREX) — involves chemically separating uranium and plutonium. A significant fraction of the plutonium stockpiles is intended to be used for MOX fuel fabrication at two industrial-scale facilities: Areva's Melox plant in Marcoule, France; and British Nuclear Group's Sellafield MOX plant in the UK. These facilities blend uranium and plutonium powders at high temperatures to create MOX pellets, which are then loaded into fuel assemblies. Researchers have recently proposed a newer method of reprocessing called uranium extraction plus (UREX+), which keeps uranium and plutonium together in the fuel cycle to avoid separating out pure plutonium. Some of the problems with reprocessing are elaborated in Chapter 4.

Interim Storage

The term "nuclear waste" technically includes six categories of waste: spent nuclear fuel, high-level waste, transuranic waste (the three most important or radioactive), as well as low-level waste, mixed waste, and uranium mill tailings. Spent nuclear fuel refers to fuel rods that have been irradiated in a nuclear reactor, meaning they contain active and short-lived fission products such as cesium and strontium as well as long-lived radionuclides. High-level waste refers to highly radioactive material resulting from the reprocessing of spent nuclear fuel, that is, the chemical processes that break down fuel rods into uranium and plutonium. Transuranic waste refers to any type of waste that has more than 100 nanocuries of alpha-emitting transuranic isotopes with half-lives greater than 20 years per gram of waste, excluding high-level waste.[16] All six types of waste must be continually tracked and monitored. Figure 5 shows an inspection of nuclear waste in the US.

To handle spent nuclear fuel and high-level waste, scientists and engineers have designed two primary types of storage: wet and dry. Wet storage

Figure 5: Workers at the US Department of Energy Inspecting Low-Level Nuclear Waste in Nevada

pools involve placing spent fuel assemblies in racks in pools of water that are made of concrete and lined with stainless steel or epoxy-based paints. The pool is enclosed within a building, and radioactivity in water is kept as low as possible.[17] The pools themselves can become quite deep, with at least three meters of water covering the top of fuel assemblies to provide radiation shielding.[18] Operators often add boron and other neutron-absorbing materials to the water to prevent the fuel from starting a chain reaction. About 90% of spent fuel globally is currently stored in wet pools.

Dry storage of spent fuel differs from wet storage, as it uses gas or air (often enhanced with helium or nitrogen to limit oxidation) instead of water as a coolant, and metal or concrete as a radiation barrier. Fuel must be stored in wet pools for several years, usually at least a decade, before they become cool enough for dry storage. Dry storage vaults use metal tubes and cylinders to store fuel, remove heat through forced air convec-tion, and have the building itself act as a radiation shield.[19] Although there are many different cask types, those in the US typically hold 20–24 pres-surized water reactor fuel assemblies, sealed in a helium atmosphere inside the cask to prevent corrosion. Decay heat is transferred by helium from the fuel to fins on the outside of the storage cask for cooling.

Permanent Sequestration of Waste

The final stage of the back end of the cycle involves the sequestration of nuclear waste. Permanent geological repositories must provide protection against every plausible scenario in which radionuclides might reach the biosphere or expose humans to dangerous levels of radiation. These risks include groundwater seeping into the repository, the corrosion of waste containers, the leaching of radionuclides, and the migration of contami-nated groundwater toward areas where it might be used as drinking water or for agriculture.

Decommissioning

The last stage of the nuclear lifecycle involves the decommissioning and dismantling of the reactor, as well as reclamation of the uranium mine site. After a cooling-off period that may last as long as 50–100 years, reactors must be dismantled and cut into small pieces to be packed in containers for final disposal. Nuclear plants have an operating lifetime of 40 years, but decommissioning usually takes at least 60 years.[20] While it will vary along with the technique and the reactor type, the total energy required for decommissioning can be as much as 50% more than the energy needed for original construction. At uranium mines, the overburden of rock covering the area must be replaced and replanted with indigenous vegetation, and radioactive tailings must be treated and contained before land can be reclaimed.

Drivers of Nuclear Power Expansion

Why have countries spent billions of dollars collaborating on nuclear technology? Of all the daunting global challenges facing the electricity sector, three seem to be most significant: the need to provide basic energy services to the world's poor; the need to find sources of energy that are less greenhouse gas-intensive; and the need to keep costs low, both for ratepayers and for governments. Proponents of nuclear power believe it is the only technology that can satisfy all three of these critical needs (although Chapter 8 raises serious doubts as to the legitimacy of these drivers).

Fighting Energy Poverty

Denying electricity and the services it can provide to those in need promotes discrimination in the vein of what is sometimes called "environmental racism." At least one billion people — roughly one sixth of the global population — have little to no access to electricity.[21] Without electricity, millions of women and children are typically forced to spend significant amounts of time searching for firewood, and then combusting wood and charcoal indoors to heat their homes or prepare meals. The health consequences alone of this combustion are monumental. Scientists estimate that indoor air pollution kills 2.8 million people every year — almost equal to the number of people dying annually from HIV/AIDS. Close to one million of these deaths (910,000, to be precise) occur in children under the age of five, who must suffer their final months of life dealing with debilitating respiratory infections, chronic obstructive pulmonary disease, and lung cancer.[22] In essence, nuclear power is seen as one of the few options that can prevent a form of "energy apartheid" whereby people in the Western world use large amounts of energy, have higher standards of living, and enjoy longer life expectancies, while those in undeveloped nations have no access to energy and die earlier.[23]

Climate Change and the Environment

Proponents of nuclear power believe that it is a much better option for generating power without releasing significant amounts of greenhouse gases or toxic pollution. Greenhouse gas emissions are expected to increase approximately 135% in the US and Canada by 2030 from today's levels under a business-as-usual scenario.[24] As Robert K. Dixon, former Head of the Technology Policy Division at the International Energy Agency (IEA), declared in 2008, "Without substantial technology and policy changes, fossil fuels will remain fuels of choice well into the future."[25]

Advocates of nuclear power have therefore framed nuclear energy as an important part of any solution aimed at fighting climate change and reducing greenhouse gas emissions. The Nuclear Energy Institute, discussing India and China, reminded the public that "it is important to influence them to build emission free sources of energy like nuclear."[26] When President George W. Bush signed the Energy Policy Act in August 2005, he remarked that "only nuclear power plants can generate massive amounts of electricity without emitting an ounce of air pollution or greenhouse gases."[27] The late Mr. Nicholas Ridley, former Secretary of State for the Environment in the UK, was even more explicit, stating on BBC television: "There is absolutely no doubt that if you want to arrest the Greenhouse Effect you should concentrate on a massive increase in nuclear generating capacity. Nuclear power stations give out no sulfur and carbon dioxide, so they are the cleanest form of power generation."[28]

Even some former nuclear power skeptics have embraced the efficacy of nuclear power as a solution to the global climate crisis. Patrick Moore, co-founder of Greenpeace and once a vocal opponent of nuclear power, has publicly stated that "nuclear energy . . . remains the only practical, safe and environmentally friendly means of reducing greenhouse gas emissions and addressing energy security."[29] Similarly, the environmentalist James Lovelock goes so far as to argue that nuclear power is one of the only options that can meet electricity demand as we transition to cleaner energy sources, and that its risks are "insignificant compared with the real threat of intolerable and lethal heat-waves" associated with climate change.[30] (Problems with this line of thinking are presented in Chapter 5.)

Also, one supposed benefit to nuclear power plants is that they produce an incredibly small volume of waste and pollution compared to fossil-fueled power stations. The impact on human health and environmental quality from fossil fuel combustion is more immediately drastic than that from nuclear power, since nuclear waste becomes less toxic with time as radioactive materials decay, whereas the chemicals emitted from fossil fuels become quickly ingested by humans and other organisms. As Hans Blix, former Director General of the International Atomic Energy Agency (IAEA), succinctly put it:

> A 1,000 megawatt equivalent (MWe) coal plant with optimal pollution abatement equipment will annually emit into the atmosphere 900 tons of sulfur dioxide, 4,500 tons of nitrous oxide, 1,300 tons of particulates, and 6.5 million tons of carbon dioxide. . . . By contrast, a nuclear plant of 1,000 MWe capacity produces annually some 35 tons of highly radioactive spent fuel.[31]

One study estimated that the waste generated by a large nuclear plant per year was 2 million times smaller by weight and 1 billion times smaller by volume than the waste from a coal-burning plant.[32] Such attributes make nuclear power attractive to some advocates wishing to produce electricity with minimal damage to the environment.[33]

Economics and Cost

Nuclear power plants do have very low historical production costs, and recently have improved their performance due to better capacity factors and operating procedures. They key word here is *historical*: many of these plants have been operating for years, and their capital costs have all but been paid off. (On the disadvantageous side, nuclear power plants have long construction lead times and are exposed to cost overruns, and future plants will likely be much more expensive than the ones already built. This is discussed in Chapter 4.)

Although particular production costs will differ by design, site requirements, and the rate of capital depreciation, the existing light water reactors — which make up a majority of the world's nuclear power fleet — produce electricity at costs between 2.5 cents/kWh and 7 cents/kWh. This makes them cheaper than almost any other source of electricity on the global market today.[34] Some coal plants and hydroelectric facilities also produce power at as cheaply as 2 cents/kWh, but most other sources of supply tend to cost more than 5 cents/kWh.

A second advantage concerns improvements in operational performance and maintenance, due largely to decades of research and development along with government subsidies. The World Nuclear Association reported that existing plants in the US, for example, improved their capacity factor by 40% from 1990 to 2008, with the average plant now operating 91.1% of the time.[35]

The industry has incorporated research findings on human factors and safety culture through groups and organizations such as the IAEA and the World Association of Nuclear Operators (WANO), created after the Chernobyl accident in 1986.[36] These efforts have produced dividends, as evidenced by one meta-survey of nuclear power plant performance that found existing plants were getting safer. The study noted that, industry-wide, 1.04 accidents occurred per 200,000 worker hours in 1990 but only 0.28 accidents occurred in 2003.[37] The study noted that better occupational safety and health regulations in Europe and North America, improved medical knowledge, and better emergency care and first aid techniques have reduced the likelihood of accidents and lessened their impact when they do occur.

Plans for Nuclear Expansion

Consequently, many believe that nuclear power is set for rapid expansion. Indeed, nuclear power plants are already being planned or constructed in the US, Europe, and Asia.

In the US, over the past two decades, nuclear power plants have been quietly but surely expanding their generating capacity. The Nuclear Regulatory Commission (NRC) approved 2,200 MW of capacity upgrades to existing nuclear plants between 1988 and 1999, and nuclear facilities are seeking approval for another 842 MW.[38] Following the 2002 unveiling of the DOE's "Nuclear Power 2010 Program," targeted at demonstrating "new regulatory processes leading to a private sector decision by 2005 to order new nuclear power plants for deployment in the United States in the 2010 timeframe," three large utilities — Exelon, Entergy, and Dominion — filed early site permits for the construction of new nuclear plants in Illinois, Texas, and Virginia, respectively.[39] The Energy Policy Act of 2005 also significantly bolstered plans for nuclear power by extending liability limits for nuclear accidents under the Price–Anderson Act for another 20 years, authorizing the construction of new DOE research reactors, and establishing hefty loan and insurance programs to make the construction of new nuclear reactors more attractive. After the Act was passed, between 2005 and 2007 the NRC received notices of application for at least 28 new nuclear units from a plethora of utilities and energy consortia, and 30 applications for new reactor units were filed by the close of 2009.[40]

In Europe, utilities operate 145 nuclear reactors throughout 15 of the 27 countries in the European Union, for a total of 131,820 MW of installed capacity which provided 31% of electricity generated in 2007.[41] France plans to replace 58 reactors with new Generation III pressurized water reactors at a rate of 1,600 MW per year. Even Ukraine, the site of the worst nuclear accident in the technology's history, is planning to construct 22 new nuclear power plants by 2030. In East and South Asia, there are 109 nuclear power reactors in operation, 18 under construction, and plans to build another 110. If one takes government proclamations at face value, 319 new nuclear power plants have been planned and proposed for a total of 325,488 MW of capacity, which would need more than 64,000 additional tons of uranium each year to operate.[42]

The fastest growth in nuclear generation is expected to occur in China, India, Japan, and South Korea. The Chinese Academy of Sciences has even embarked on an ambitious public relations campaign to make nuclear power more popular. Chinese operators already have five units under construction and 50 proposed by 2020, and they plan to quadruple nuclear capacity from 7.6 GW in 2008 to more than 40 GW by 2020. India, which meets only 3% of electricity demand with nuclear power, planned a tenfold increase from 700 MW to 7280 MW by 2010 (though this has not yet occurred). Japan, which currently operates 55 commercial light water reactors, is seeking to increase its share of nuclear electricity from about 30% in 2008 to 40% over the next two decades. Japanese utilities thus have two plants under construction and 11 more planned. South Korea, which currently operates 16 reactors, has six under construction and eight more planned by 2015, implying a 100% increase in nuclear power generation.

Even developing countries in Southeast Asia are attempting to embrace nuclear power. Under a Regional Cooperative Agreement signed in 1972, Australia, Bangladesh, China, India, Indonesia, Japan, Malaysia, Mongolia, Myanmar, New Zealand, Pakistan, the Philippines, Singapore, South Korea, Sri Lanka, Thailand, and Vietnam have agreed to promote cooperative research and training in nuclear-related fields.[43] Vietnam is aiming for its first nuclear plant by 2015; Malaysia has plans for its first nuclear power plant by 2020; Thailand is planning to install 4 GW of nuclear capacity by 2020; and Indonesia's 4-GW Mount Muria plant is currently on hold, but is scheduled to start construction in 2011 and become operational by 2018. In other parts of the world, 30 nuclear plants are being built in 12 countries, with additional units in the planning stages for Argentina, Brazil, the Czech Republic, Finland, France, Mexico, Peru, Romania, and Russia.

As the chapters to follow show, however, many of these planned and proposed nuclear power plants will never be built. Despite all of the recent efforts to research, design, plan, construct, operate, and upgrade nuclear power plants, transitioning to an energy economy based on significant expansions in nuclear power would bring serious consequences. The remaining chapters in the book will document how nuclear power plants create massive external costs that are not subsumed by ratepayers or even present generations. Nuclear facilities rely almost entirely on government subsidies for construction, storage, and liability. While, historically, the costs of nuclear power plants appear to be low, in the near future the cost of building new nuclear plants will be outrageously high, and the promise of Generation IV reactors is entirely theoretical and will require billions of dollars in further research before the industry can construct even an experimental reactor.

In the end, a nuclear reactor, despite all of its complexities, is merely another way to boil water to generate electricity. Relying on nuclear reactions, uranium mines, enrichment facilities, and the like is an incredibly complicated way to do what other technologies accomplish by harnessing the power of wind or the kinetic energy of falling water. One commentator joked that relying on nuclear energy to meet the growing demand for electricity is akin to "smoking cigarettes to keep the weight off."[44] Or, as Christian Parenti put it:

> Nuclear fission is a mind-bogglingly complex process, a sublime, truly Promethean technology. Let's recall: it involves smashing a subatomic particle, a neutron, into an atom of uranium-235 to release energy and more neutrons, which then smash other atoms that release more energy and so on infinitely, except the whole process is controlled and used to boil water, which spins a turbine that generates electricity. . . . In this nether realm, where industry and science seek to reproduce a process akin to that which occurs inside the sun, even basic tasks — like moving the fuel rods, changing spare parts — become complicated, mechanized and expensive. Atom-smashing is to coal power, or a windmill, as a Formula One race-car engine is to the mechanics of a bicycle.[45]

As we shall also see in Chapter 7, many advantages exist by relying on the bicycle.

Endnotes

1   Norman Metzger, *Energy: The Continuing Crisis* (New York: Thomas Y. Crowell Company, 1984).

2   Modified from World Nuclear Association, "World Nuclear Power Reactors and Uranium Requirements," April 1, 2010.

3   Source: Antony Froggatt, "Nuclear Self-Sufficiency — Can Nuclear Power Pave the Road Towards Energy Independence?," Presentation to the "Towards a Nuclear Power Renaissance" Conference in Potsdam, Germany, March 4–5, 2010.

4   Victor Gilinsky, Marvin Miller, and Harmon Hubbard, *A Fresh Examination of the Proliferation Dangers of Light Water Reactors* (Washington, D.C.:

Nonproliferation Policy Education Center, October 22, 2004). Gilinsky *et al.*, interestingly, also note: "In the 1950s, before the advent of nuclear power plants, the United States tried to control the uranium market by buying up uranium at high prices. This naturally encouraged exploration that demonstrated that uranium was plentiful and negated the US effort at control. With easy access to uranium but lacking indigenous uranium enrichment facilities, Britain, France, and Canada opted for reactor designs that utilized natural uranium fuel and heavy water or graphite as the neutron moderator. In the late 1950s and early 1960s, they interested Italy, Japan, India, and other countries in heading in this direction. Not only did this threaten America's competitive position but it also threatened to spread a type of reactor that lent itself easily to production of plutonium. In fact the first British and French power reactors were based on their military plutonium production reactors. America's advantage was two-fold. The United States had developed a compact, and therefore relatively low-cost, LWR [light water reactor] design based on a naval propulsion reactor design. And the United States had invested heavily in gaseous diffusion plants in Tennessee, Kentucky, and Ohio to enrich uranium for weapons. The LWR could only operate on enriched uranium, that is, uranium more concentrated in the active uranium-235 isotope than natural uranium. By virtue of its huge enrichment capacity, the United States had an effective monopoly on the production of this fuel. Moreover, as the cost of the plants had been largely assigned to the military budget, the United States decided to sell the stuff at low prices that did not defray the massive investment. It was a price that at the time no other country could even hope to offer in the future. From the point of view of customers, it was a deal that was hard to refuse, even if it came with US control conditions. Ultimately, the amount of engineering invested in these designs and the depth of experience with them overwhelmed any conceptual advantages other reactor types may have had. While not the exclusive choice — Canada and India continued developing the natural uranium/heavy water designs that evolved into the CANDU reactor — the LWR became the standard reactor type around the world. In the late 1960s France switched to LWRs, and Britain did later. Other European manufacturers in Germany and Sweden chose LWRs. The Soviets eventually did, too."

[5] Massachusetts Institute of Technology, *The Future of Nuclear Power: An Interdisciplinary MIT Study* (Cambridge, MA: MIT, 2003).

[6] Yury Yudin, *Multilateralization of the Nuclear Fuel Cycle: Assessing the Existing Proposals* (Geneva: United Nations Institute for Disarmament Research, 2009).

[7] Source: *Ibid.*

[8] International Institute for Strategic Studies, *Preventing Nuclear Dangers in Southeast Asia and Australasia* (London: IISS, September 2009).

[9] See Scott W. Heaberlin, *A Case For Nuclear Generated Electricity* (Columbus, OH: Batelle Press, 2003); and David Fleming, *The Lean Guide to Nuclear Energy: A Lifecycle in Trouble* (London: The Lean Economy Connection, 2007).

[10] Yudin (2009), p. 67.

[11] Vasilis Fthenakis and Hyung Chul Kim, "Greenhouse-Gas Emissions from Solar Electric- and Nuclear Power: A Life-Cycle Study," *Energy Policy* 35 (2007), pp. 2549–2557.

[12] Benjamin K. Sovacool and Christopher Cooper, "Nuclear Nonsense: Why Nuclear Power Is No Answer to Climate Change and the World's Post-Kyoto Energy Challenges," *William & Mary Environmental Law and Policy Review* 33(1) (Fall, 2008), pp. 1–119.

[13] Scott W. White, *Energy Balance and Lifetime Emissions from Fusion, Fission, and Coal Generated Electricity* (Madison, WI: M.S. thesis, University of Wisconsin, 1995, UWFDM-993).

[14] Sovacool and Cooper (2008).

[15] Constance Perin, "Operating as Experimenting: Synthesizing Engineering and Scientific Values in Nuclear Power Production," *Science, Technology, & Human Values* 23(1) (Winter, 1998), pp. 98–128.

[16] Richard B. Stewart, "U.S. Nuclear Waste Law and Policy: Fixing a Bankrupt System," *New York University Environmental Law Journal* 17 (2008), pp. 783–825.

[17] Matthew Bunn, John P. Holdren, Allison Macfarlane, Susan E. Pickett, Atsuyuki Suzuki, Tatsujiro Suzuki, and Jennifer Weeks, *Interim Storage of Spent Nuclear Fuel: A Safe, Flexible, and Cost-Effective Near-Term Approach to Spent Fuel Management* (Cambridge, MA and Tokyo: A Joint Report from the Harvard University Project on Managing the Atom and the University of Tokyo Project on Sociotechnics of Nuclear Energy, June 2001).

[18] Allison Macfarlane, "Interim Storage of Spent Fuel in the United States," *Annual Review of Energy and Environment* 26 (2001), pp. 201–235.

[19] Bunn *et al.* (2001).

[20] J.L.R. Proops, Philip W. Gay, Stefan Speck, and Thomas Schroder, "The Lifetime Pollution Implications of Various Types of Electricity Generation," *Energy Policy* 24(3) (1996), pp. 229–237.

[21] Just 64% of the population in developing countries as a whole have access to electricity. In Asia and Africa, the numbers are even lower: 40.8% for South and Southeast Asia, 34.3% for Africa, and 22.6% for Sub-Saharan Africa. See International Energy Agency, *World Energy Outlook 2009* (Paris: OECD, 2009).

[22] John P. Holdren and Kirk R. Smith, "Energy, the Environment, and Health," in Tord Kjellstrom, David Streets, and Xiadong Wang (eds.), *World Energy Assessment: Energy and the Challenge of Sustainability* (New York: United Nations Development Programme, 2000), pp. 61–110.

[23] Denis E. Beller, "Atomic Time Machines: Back to the Nuclear Future," *Journal of Land Resources and Environmental Law* 24 (2004), pp. 41–61.

[24] International Energy Agency (2009).

[25] Quoted in Benjamin K. Sovacool, Hans H. Lindboe, and Ole Odgaard, "Is the Danish Wind Energy Model Replicable for Other Countries?," *Electricity Journal* 21(2) (2008), p. 29.

[26] Quoted in Sovacool and Cooper (2008), p. 17.

[27] *Ibid.*, p. 18.

[28] *Ibid.*

[29] *Ibid.*

[30] J. Lovelock, *The Revenge of Gaia: Earth's Climate Crisis & the Fate of Humanity* (New York: Basic Books, 2003).

[31] Hans Blix, "Nuclear Energy in the 21st Century," *Nuclear News* (September, 2001), pp. 34–48.

[32] B.L. Cohen, "Perspectives on the High Level Waste Disposal Problem," *Interdisciplinary Science Reviews* 23 (1998), pp. 193–203.

[33] S. Pacala and R. Socolow, "Stabilization Wedges: Solving the Climate Problem for the Next 50 Years with Current Technologies," *Science* 305 (August 13, 2004), pp. 968–972.

[34] Quirin Schiermeier, Jeff Tollefson, Tony Scully, Alexandra Witze, and Oliver Morton, "Electricity Without Carbon," *Nature* 454 (August, 2008), pp. 816–823.

[35] World Nuclear Association, "Nuclear Power in the USA," November 12, 2009, available at http://www.world-nuclear.org/info/inf41.html/.

[36] M.V. Ramana, "Nuclear Power: Economic, Safety, Health, and Environmental Issues of Near-Term Technologies," *Annual Review of Environment and Resources* 34 (2009), pp. 127–152.

[37] L.C. Cadwallader, *Occupational Safety Review of High Technology Facilities* (Idaho Falls, ID: Idaho National Engineering and Environmental Laboratory, January 2005, INEEL/EXT-05-02616).

[38] Neil J. Numark and Robert D. MacDougall, "Nuclear Power in Deregulated Markets: Performance to Date and Prospects for the Future," *Tulane Environmental Law Journal* 14 (2001), pp. 465–466.

[39] Thomas B. Cochran, Director of the Natural Resources Defense Council's Nuclear Program, "The Future Role of Nuclear Power in the United States," Presentation to the Western Governors' Association North American Energy Summit (April 15, 2004).

[40] Paul W. Benson and Fred Adair, "Nuclear Revolution: How to Ease the Coming Upheaval in the Nuclear Power Industry," *Public Utilities Fortnightly* (July, 2008).

[41] D. Haas and D.J. Hamilton, "Fuel Cycle Strategies and Plutonium Management in Europe," *Progress in Nuclear Energy* 49 (2007), p. 575.

[42] Andrew Symon, "Southeast Asia's Nuclear Power Thrust: Putting ASEAN's Effectiveness to the Test?," *Contemporary Southeast Asia* 30 (2008), p. 123.

[43] A.P. Jayaraman, "Nuclear Energy in Asia," Presentation to the Seminar on Sustainable Development and Energy Security (April 22–23, 2008), p. 13.

[44] Gabriel Walt, "Is Nuclear Power a Solution?," wattwatt.com, August 3, 2007.

[45] Christian Parenti, "Nuclear Power Is Risky and Expensive," in Peggy Becker (ed.), *Alternative Energy* (New York: Gale, 2010), pp. 50–55.

3

Safety and Reliability: Dealing with "Normal Accidents"

An ironic moment occurred on March 31, 1979. That evening, then-US Secretary of Energy James Schlesinger was testifying before the American Congress on ways to expedite the licensing process for nuclear reactors, arguing that onerous requirements were no longer needed given the inherent safety of new designs. At the same time, the Nuclear Regulatory Commission (NRC) Chairman Joe Hendrie was transmitting evacuation orders to Governor Richard L. Thornburgh in Pennsylvania because of the accident at Three Mile Island (TMI). Unknown to Schlesinger, the NRC had long suspected that an accident would occur at TMI, previously ordering the shutdown of five similarly designed nuclear power plants based on errors discovered in a computer program used to assess the stresses on power plant pipes and cooling systems during an earthquake. A few days before the accident in March 1979, NRC inspectors had even warned the Commissioner that the TMI design was unsafe and should be shut down immediately. The NRC was in the process of considering what to do when the accident occurred.[1]

The story does not end there. Rather than admit to the inherent flaws with their reactor designs, the nuclear industry ran a sleek public relations campaign a few months after the accident featuring the physicist Edward Teller in newspaper and television advertisements. In these advertisements,

Teller solemnly told viewers (or, in newspaper versions, expressed in very large bold-faced type) that "I was the only victim of Three Mile Island." Even though Teller was nowhere near Pennsylvania at the time of the accident, he claimed that he suffered a heart attack a few weeks later because he had been working tirelessly to refute senseless anti-nuclear propaganda.[2]

The lessons from this story are numerous and possibly prophetic. It reveals that various organizations promoting nuclear power do not always share information and can make mistakes (as in the Secretary of Energy believing designs to be safe when the NRC did not). It shows that some scientists and engineers involved in the industry, such as Teller, have optimistic views about atomic energy and intolerance for skepticism. It demonstrates that nuclear reactors are extremely dangerous when they malfunction. It also implies that the nuclear industry will utilize public opinion and savvy media techniques to insulate itself from criticism.

This chapter explores the safety and reliability concerns with existing and new nuclear power plants. It looks at the historical record of incidents and accidents, current risks with the global reactor fleet, and future risks with new reactors. It also explores an often-ignored component, namely the scarcity of high-quality materials and skilled labor to build and operate nuclear units, and finally discusses the technical challenges related to finding high-quality uranium fuel and a declining energy payback ratio.

Safety and Accidents

While the Chair of the Public Information Committee of the American Nuclear Society has publicly stated that "the industry has proven itself to be the safest major source of electricity in the Western world,"[3] the history of nuclear power proves otherwise. The safety record of nuclear plants is lackluster at best. For one salient example, consider that Ukraine still has a Ministry of Emergency, some 24 years after the Chernobyl nuclear disaster warranted its creation. This section focuses on historical accidents at nuclear facilities, with a special emphasis on two of the most famous accidents at Chernobyl and TMI, as well as the risk of future accidents.

Whenever one talks about safety culture, nuclear accidents, and reliability, it is important to be clear about the terms. Part of the confusion stems from how one defines an accident. The NRC and the nuclear community generally separate unplanned events into two classes: incidents and accidents. Incidents are unforeseen events and equipment failures that occur during normal plant operation, resulting in no offsite releases of radiation or severe damage to equipment; accidents refer to either offsite releases of radiation or severe damage to plant equipment.[4] The International Nuclear and Radiological Event Scale communicates the significance of a nuclear and radiological event through a ranking system of seven levels: levels 1–3 are "incidents" while levels 4–7 are "accidents," with a "level 7 major accident" consisting of "a major release of radioactive material with widespread health and environmental effects requiring implementation of planned and extended countermeasures."[5] The Paul Scherrer Institute manages an Energy-Related Severe Accident Database (ENSAD), which takes a slightly different approach. For the ENSAD, a severe accident is one which involves one of the following: at least 5 fatalities, at least 10 injuries, 200 evacuees, 10,000 tons of hydrocarbons released, more than 25 km² of cleanup, or more than US$5 million in economic losses.[6]

Under the classifications of accidents from the NRC, the International Nuclear and Radiological Event Scale, and even the ENSAD, the number of nuclear accidents is low. However, if one redefines an accident to be an incident that results in either the loss of human life or more than US$50,000 of property damage, a very different picture emerges. One study identified no less than 76 nuclear accidents meeting this definition, totaling more than US$19 billion in damages worldwide, from 1947 to 2008.[7] These accidents accounted for 41% of all accident-related property damages globally. Such accidents involved meltdowns, explosions, fires, and loss of coolant, and occurred during both normal operation and extreme emergency conditions (such as droughts and earthquakes). Another index of nuclear power accidents that included costs beyond death and property damage — such as injury to or irradiation of workers and malfunctions that did not result in shutdowns or leaks — documented 956 incidents from 1942 to 2007.[8] Yet another study documented that, between the 1979 accident at TMI and 2009, there were more than 30,000 mishaps at US nuclear power plants alone, many with the potential to have caused serious meltdowns.[9] Researchers at American University even calculated at least

124 "hazardous incidents" at nuclear units in India between 1993 and 1995.[10] The 200 nuclear facilities in France, including power plants, uranium enrichment and conversion plants, reprocessing plants, fuel fabrication plants, surface repositories for waste, and experimental sites for geologic disposal, declare in total 700–800 serious incidents or significant safety events each year.[11]

One of the first accident studies, conducted by the US Atomic Energy Commission in 1975, looked at the performance of early nuclear plants in terms of occupational injury and death over 32 years of development. They documented 111 accidents involving unplanned releases of radioactivity that exposed 317 people to excess radiation as high as 80,000 rads ("safe" levels are fiercely debated, but are generally less than 10 rads). The study described 321 total fatalities, of which 184 occurred during construction, 212 during operations, and 16 during inspections and government functions (the sums do not match, as one fatality could fall into multiple categories), along with a total of 19,225 injuries not involving radiation for an unusually high frequency rate of 2.75 injuries per million man-hours.[12] Such incidents and accidents not only harm human beings, but also take their toll on operating performance. Using data from US, French, Belgian, German, Swedish, and Swiss nuclear power plants, one study found mean durations of continual operation from 35 to 88 days, meaning these plants saw scores of unplanned outages, half of which were related to equipment failure.[13] Adato *et al.* also cited more than 200 serious accidents and partial meltdowns in commercial nuclear power plants from 1960 to 1980 in the US.[14] Even the Paul Scherrer Institute's ENSAD, despite defining accidents differently, suggested that the latent effects of the Chernobyl disaster make nuclear power 41 times more dangerous than equivalently sized coal, oil, natural gas, and hydroelectric projects.[15]

The above figures tend to be conservative, as they frequently do not include accidents and incidents at research reactors and other parts of the nuclear fuel chain. Mistakes are not limited to reactor sites. For example, accidents at the Savannah River reprocessing plant have already released 10 times as much radioiodine than the accident at TMI; and a fire at the Gulf United plutonium facility in New York in 1972 scattered an undisclosed amount of plutonium into residential neighborhoods, forcing the plant to shut down permanently.[16] A similar fire at the Rocky Flats reprocessing plant in Colorado released hundreds of pounds of plutonium oxide dust into the surrounding environment. When United Nuclear Corporation's uranium mine tailings dam near Church Rock, New Mexico, burst in July 1979, it released 93 million gallons of radioactive water and 1,000 tons of radioactive sediment into local rivers. Outside of military weapons testing, this accident remains the single largest release of radioactive materials in the US. Almost 2,000 Navajo were directly affected with undrinkable water, while sheep and livestock were heavily contaminated with lead-210, polonium-210, thorium-230, and radium-236.[17] At the Mayak Industrial Reprocessing Complex in the Southern Urals, Russia, the overheating of a storage tank with nitrate acetate salts exploded in 1957, releasing a massive amount of radioactive material over 20,000 km^2 in Chelyabinsk and Sverdlovsk, causing the evacuation of 272,000 people. In September 1994, an explosion at the Serpong research reactor in Indonesia was triggered by the ignition of methane gas that had seeped from packages being removed from a laboratory storage room, which exploded when a worker lit a cigarette.[18]

Accidents have also occurred when nuclear reactors are shut down to be refueled or when fuel is to be transitioned into storage. In 1999, operators were beginning to load spent fuel into dry storage at the Trojan Reactor in Oregon when they found that the zinc-carbon coating intended to protect against borated water had started producing hydrogen, causing a small explosion. Similar hydrogen explosions have occurred at the Palisades plant in Michigan and the Point Beach reactor in Wisconsin, when operators were trying to weld casks shut. Follow-up investigations identified poor quality assessment, not following procedures, and failure to document previous repairs to casks as the likely causes.[19]

Onsite accidents at nuclear reactors and fuel facilities, unfortunately, are not the only cause of concern. The August 2003 blackout on the US East Coast revealed that more than a dozen nuclear reactors in the US and Canada were not properly maintaining backup diesel generators. In Ontario, during the blackout, reactors designed to automatically unlink from the grid and remain in standby mode instead went into full automatic shutdown, with only two of 12 reactors shutting down as planned. Because they must connect to another source of electricity to keep coolant circulating, all nuclear facilities maintain several backup diesel generators onsite for use in the event of a power loss. From September 2002 to August 2003, plant operators declared emergency diesel generators inoperable in 15 reported instances. In seven of those cases, a complete shutdown of the plant was required; and on four of those occasions, all backup generators failed at the same time. In April 2003, the Cook nuclear plant in Western Michigan shut down when emergency water flow to all four diesel generators was blocked by an influx of fish on cooling-system intake screens. These examples suggest that relying on backup systems to respond to blackouts presents a great likelihood of failure and can themselves create dangerous situations. More worryingly, since spent fuel ponds do not receive backup power from emergency diesel generators, when offsite power goes down, pool water cannot be recirculated to prevent boiling, evaporation, and exposure of fuel rods; the result is an increased risk of pool fires and explosions.[20]

Even research facilities have their own set of safety problems. Operators at the RA-2 Facility in Constituyentes, Argentina, mistakenly placed two fuel elements in the same graphite reflector, causing a criticality excursion that killed one person and injured two others. The Henry L. Stimson Center has documented numerous criticality accidents at research reactors to date, including 11 loss-of-flow accidents, 6 loss-of-cooling accidents, 25 erroneous handlings or failures of equipment, and 2 special events that have so far resulted in 21 deaths spread across the US, the Soviet Union, Japan, Argentina, and Yugoslavia.[21] The nonpartisan Government Accountability Office (GAO) recently found that 31 research facilities with reactors or nuclear materials were operating in the US for extended periods of time in noncompliance with nuclear safety licensing requirements.[22] The GAO concluded that:

> The Department of Energy has structured its independent oversight office, the Office of Health, Safety, and Security (HSS), in a way that falls short of meeting our key elements of effective independent oversight of nuclear safety. . . . HSS falls short of fully meeting our five key elements of effective oversight of nuclear safety: independence, technical expertise, ability to perform reviews and require that its findings are addressed, enforcement authority, and public access. First, we found that HSS has no role in reviewing the safety basis for new high-hazard nuclear facilities, no routine site presence, and its head is not comparable in rank to the program office

heads. Second, HSS does not have some technical expertise in nuclear safety review and has vacancies in critical nuclear safety positions. Third, HSS lacks basic information about nuclear facilities, has gaps in its site inspection schedule, and does not routinely ensure that its findings are effectively addressed. Fourth, HSS enforcement actions have not prevented some recurring nuclear safety violations. Finally, HSS restricts public access to nuclear safety information.[23]

Such trends are worrying, to say the least, as the national laboratories in the US are often prized for having highly trained nuclear specialists. If these specialists cannot conform to safety standards, it raises serious questions about how operators and researchers in other countries can.

The author's own compilation reveals 99 nuclear accidents totaling US$20.5 billion in damages worldwide from 1952 to early 2010 (see Table 1). Looking at only relatively recent accidents, these numbers translate to more than one incident and US$330 million in damages every year for the past three decades. When compared to fatalities from other energy sources, nuclear power ranks as the second most fatal source of energy supply (after hydroelectric dams) and is ranked higher than oil, coal, and natural gas systems. Fifty-seven accidents have occurred since the Chernobyl disaster in 1986; and almost two-thirds (56 out of 99) of all nuclear accidents have occurred in the US, refuting the notion that accidents are relegated to the past or to countries without America's modern technologies or industry oversight. While only a few accidents globally involved fatalities, those that did collectively killed more people than have died in commercial US airline accidents since 1982.

Some of these accidents would be laughable if not for their seriousness, and include:

- A maintenance worker at the North Anna nuclear plant in Virginia cleaning the floor in an auxiliary building who caught his shirt on a circuit breaker, tripping the reactor and causing a four-day shutdown;
- An employee changing a light bulb in a control panel at Rancho Seco in California who accidentally dropped it into the reactor, short-circuiting sensor arrays and leading to an increase in pressure that almost cracked the reactor vessel;

Table 1: 99 Major Nuclear Power Accidents from 1952 to 2010[24]

Date	Location	Description	Fatalities	Cost (in US$ million (2006))
December 12, 1952	Chalk River, Ontario, Canada	Hydrogen explosion damages reactor interior, releasing 30 kg of uranium oxide particles	0	$45
October 8, 1957	Windscale, United Kingdom	Fire ignites plutonium piles and destroys surrounding dairy farms	33	$78
May 24, 1958	Chalk River, Ontario, Canada	Fuel rod catches fire and contaminates half of the facility	0	$67
July 26, 1959	Simi Valley, California, United States	Partial core meltdown takes place at the Santa Susana Field Laboratory's Sodium Reactor Experiment	0	$32
January 3, 1961	Idaho Falls, Idaho, United States	Explosion at the National Reactor Testing Station	3	$22
October 5, 1966	Monroe, Michigan, United States	Sodium cooling system malfunctions at the Enrico Fermi demonstration breeder reactor, causing partial core meltdown	0	$19
May 2, 1967	Dumfries and Galloway, Scotland	Fuel rod catches fire and causes partial meltdown at the Chaplecross Magnox nuclear power station	0	$76
January 21, 1969	Lucens, Canton of Vaud, Switzerland	Coolant system malfunctions at an underground experimental reactor	0	$22
May 1, 1969	Stockholm, Sweden	Malfunctioning valve causes flooding in the Agesta pressurized heavy water nuclear reactor, short-circuiting control functions	0	$14

(Continued)

Table 1: (*Continued*)

Date	Location	Description	Fatalities	Cost (in US$ million (2006))
July 16, 1971	Cordova, Illinois, United States	An electrician is electrocuted by a live cable at the Quad Cities Unit 1 reactor on the Mississippi River	1	$1
August 11, 1973	Palisades, Michigan, United States	Steam generator leak causes manual shutdown of pressurized water reactor operated by the Consumers Power Company	0	$10
March 22, 1975	Browns Ferry, Alabama, United States	Fire burns for seven hours and damages more than 1,600 control cables for three nuclear reactors, disabling core cooling systems	0	$240
November 5, 1975	Brownsville, Nebraska, United States	Hydrogen gas explosion damages the Cooper Nuclear Facility's boiling water reactor and an auxiliary building	0	$13
February 22, 1977	Jaslovske Bohunice, Czechoslovakia	Mechanical failure during fuel loading causes severe corrosion of reactor and release of radioactivity into the plant area, necessitating total decommission	0	$1,700
June 10, 1977	Waterford, Connecticut, United States	Hydrogen gas explosion damages three buildings and forces shutdown of the Millstone-1 pressurized water reactor	0	$15
February 4, 1979	Surry, Virginia, United States	Virginia Electric Power Company manually shuts down Surry Unit 2 in response to replace failed tube bundles in steam generators	0	$12

(*Continued*)

Table 1: *(Continued)*

Date	Location	Description	Fatalities	Cost (in US$ million (2006))
March 28, 1979	Middletown, Pennsylvania, United States	Equipment failures and operator error contribute to loss of coolant and partial core meltdown at the Three Mile Island nuclear reactor	0	$2,400
July 25, 1979	Saclay, France	Radioactive fluids escape into drains designed for ordinary waste, seeping into the local watershed at the Saclay BL3 Reactor	0	$5
September 12, 1979	Mihama, Japan	Fuel rods at the Mihama Nuclear Power Plant unexpectedly bow and damage the fuel supply system	0	$11
March 13, 1980	Loir-et-Cher, France	A malfunctioning cooling system fuses fuel elements together at the Saint Laurent A2 reactor, ruining the fuel assembly and forcing an extended shutdown	0	$22
November 22, 1980	San Onofre, California, United States	A worker cleaning breaker cubicles at the San Onofre pressurized water reactor contacts an energized line, electrocuting him to death	1	$1
February 11, 1981	Florida City, Florida, United States	Florida Power & Light manually shuts down Turkey Point Unit 3 after steam generator tubes degrade and fail	0	$2
March 8, 1981	Tsuruga, Japan	278 workers are exposed to excessive levels of radiation during repairs of the Tsuruga nuclear plant	0	$3
February 26, 1982	San Clemente, California, United States	Southern California Company shuts down San Onofre Unit 1 out of concerns for an earthquake	0	$1

(Continued)

Table 1: *(Continued)*

Date	Location	Description	Fatalities	Cost (in US$ million (2006))
March 20, 1982	Lycoming, New York, United States	Recirculation system piping fails at Nine Mile Point Unit 1, forcing a 2-year shutdown	0	$45
March 25, 1982	Buchanan, New York, United States	Multiple water and coolant leaks cause damage to steam generator tubes and main generator, forcing the New York Power Authority to shut down Indian Point Unit 3 for more than one year	0	$56
June 18, 1982	Seneca, South Carolina, United States	Feedwater heat extraction line fails at the Oconee 2 pressurized water reactor, damaging the thermal cooling system	0	$10
February 12, 1983	Fork River, New Jersey, United States	Oyster Creek nuclear plant fails safety inspection and is forced to shut down for repairs	0	$32
February 26, 1983	Pierce, Florida, United States	Workers discover a damaged thermal shield and core barrel support at St. Lucie Unit 1, necessitating a 13-month shutdown	0	$54
September 7, 1983	Athens, Alabama, United States	Tennessee Valley Authority discovers extensive damage to the recirculation system pipeline, requiring an extended shutdown	0	$34
September 23, 1983	Buenos Aires, Argentina	Operator error during fuel plate reconfiguration causes meltdown in an experimental test reactor	1	$65

(Continued)

Table 1: (*Continued*)

Date	Location	Description	Fatalities	Cost (in US$ million (2006))
December 10, 1983	Plymouth, Massachusetts, United States	Recirculation system piping cracks and forces the Pilgrim nuclear reactor to shut down	0	$4
April 14, 1984	Bugey, France	Electrical cables fail at the command center of the Bugey nuclear power plant and force a complete shutdown of one reactor	0	$2
April 18, 1984	Delta, Pennsylvania, United States	Philadelphia Electric Company shuts down Peach Bottom Unit 2 due to extensive recirculation system and equipment damage	0	$18
June 13, 1984	Platteville, Colorado, United States	Moisture intrusion causes 6 fuel rods to fail at the Fort St. Vrain nuclear plant, requiring an emergency shutdown from the Public Service Company of Colorado	0	$22
September 15, 1984	Athens, Alabama, United States	Safety violations, operator error, and design problems force a 6-year outage at Browns Ferry Unit 2	0	$110
March 9, 1985	Athens, Alabama, United States	Instrumentation systems malfunction during startup, convincing the Tennessee Valley Authority to suspend operations at all three Browns Ferry units	0	$1,830
June 9, 1985	Oak Harbor, Ohio, United States	Loss of feedwater provokes the Toledo Edison Company to inspect the Davis-Besse facility, where inspectors discover corroded reactor coolant pumps and shafts	0	$23

(*Continued*)

Table 1: *(Continued)*

Date	Location	Description	Fatalities	Cost (in US$ million (2006))
August 22, 1985	Soddy-Daisy, Tennessee, United States	Tennessee Valley Authority Sequoyah Units 1 and 2 fail NRC inspection due to failed silicon rubber insulation, forcing a 3-year shutdown, followed by water circulation problems that expose workers to excessive levels of radiation	0	$35
December 26, 1985	Clay Station, California, United States	Safety and control systems unexpectedly fail at the Rancho Seco nuclear reactor, ultimately leading to the premature closure of the plant	0	$672
April 11, 1986	Plymouth, Massachusetts, United States	Recurring equipment problems with instrumentation, vacuum breakers, instrument air system, and main transformer force an emergency shutdown of Boston Edison's Pilgrim nuclear facility	0	$1,001
April 26, 1986	Kiev, Ukraine	Mishandled reactor safety test at the Chernobyl nuclear reactor causes steam explosion and meltdown, necessitating the evacuation of 300,000 people from Kiev and dispersing radioactive material across Europe	4,056	$6,700
May 4, 1986	Hamm-Uentrop, Germany	Operator actions to dislodge a damaged fuel rod at an experimental high-temperature gas reactor release excessive radiation to 4 km^2 surrounding the facility	0	$267

(Continued)

Table 1: (*Continued*)

Date	Location	Description	Fatalities	Cost (in US$ million (2006))
May 22, 1986	Normandy, France	A reprocessing plant at Le Hague malfunctions, exposing workers to unsafe levels of radiation and forcing five to be hospitalized	0	$5
March 31, 1987	Delta, Pennsylvania, United States	Philadelphia Electric Company shuts down Peach Bottom Units 2 and 3 due to cooling malfunctions and unexplained equipment problems	0	$400
April 12, 1987	Tricastin, France	Areva's Tricastin fast breeder reactor leaks coolant, sodium, and uranium hexachloride, injuring seven workers and contaminating water supplies	0	$50
May 4, 1987	Kalpakkam, India	Fast breeder test reactor at Kalpakkam has to shut down due to the simultaneous occurrence of pump failures, faulty instrument signals, and turbine malfunctions that culminate in a refueling accident that ruptures the reactor core with 23 fuel assemblies, resulting in a 2-year shutdown	0	$300
July 15, 1987	Burlington, Kansas, United States	A safety inspector dies from electrocution after contacting a mislabeled wire	1	$1
December 17, 1987	Hesse, Germany	Stop valve fails at the Biblis Nuclear Power Plant and contaminates the local area	0	$13

(*Continued*)

Table 1: (*Continued*)

Date	Location	Description	Fatalities	Cost (in US$ million (2006))
December 19, 1987	Lycoming, New York, United States	Fuel rod, waste storage, and water pumping malfunctions force the Niagara Mohawk Power Corporation to shut down Nine Mile Point Unit 1	0	$150
March 29, 1988	Burlington, Kansas, United States	A worker falls through an unmarked manhole and electrocutes himself when trying to escape	1	$1
September 10, 1988	Surry, Virginia, United States	Refueling cavity seal fails and destroys the internal pipe system at Virginia Electric Power Company's Surry Unit 2, forcing a 12-month outage	0	$9
March 5, 1989	Tonopah, Arizona, United States	Atmospheric dump valves fail at Arizona Public Service Company's Palo Verde Unit 1, leading to a main transformer fire and emergency shutdown	0	$14
March 17, 1989	Lusby, Maryland, United States	Inspections at Baltimore Gas & Electric's Calvert Cliff Units 1 and 2 reveal cracks at pressurized heater sleeves, forcing extended shutdowns	0	$120
September 10, 1989	Tarapur, Maharashtra, India	Operators at the Tarapur nuclear power plant discover that the reactor had been leaking radioactive iodine through its cooling structures and discover radiation levels of iodine-129 more than 700 times the normal level; repairs to the reactor take more than one year	0	$78

(*Continued*)

Table 1: (*Continued*)

Date	Location	Description	Fatalities	Cost (in US$ million (2006))
November 24, 1989	Greifswald, East Germany	Electrical error causes a fire in the main trough that destroys control lines and 5 main coolant pumps, and almost induces a meltdown	0	$443
November 17, 1991	Scriba, New York, United States	Safety and fire problems force the New York Power Authority to shut down the FitzPatrick nuclear reactor for 13 months	0	$5
April 21, 1992	Southport, North Carolina, United States	NRC forces the Carolina Power & Light Company to shut down Brunswick Units 1 and 2 after emergency diesel generators fail	0	$2
May 13, 1992	Tarapur, Maharashtra, India	A malfunctioning tube causes the Tarapur nuclear reactor to release 12 curies of radioactivity	0	$2
February 3, 1993	Bay City, Texas, United States	Auxiliary feedwater pumps fail at South Texas Project Units 1 and 2, prompting a rapid shutdown of both reactors	0	$3
February 27, 1993	Buchanan, New York, United States	New York Power Authority shuts down Indian Point Unit 3 after the AMSAC system fails	0	$2
March 2, 1993	Soddy-Daisy, Tennessee, United States	Equipment failures and broken pipes cause the Tennessee Valley Authority to shut down Sequoyah Unit 1	0	$3

(*Continued*)

Table 1: *(Continued)*

Date	Location	Description	Fatalities	Cost (in US$ million (2006))
March 31, 1993	Bulandshahr, Uttar Pradesh, India	The Narora Atomic Power Station suffers a fire at two of its steam turbine blades, damaging the heavy water reactor and almost leading to a meltdown	0	$220
December 25, 1993	Newport, Michigan, United States	Detroit Edison Company is prompted to shut down Fermi Unit 2 after the main turbine experiences catastrophic failure due to improper maintenance	0	$67
April 6, 1994	Tomsk, Russia	Pressure buildup causes mechanical failure at the Tomsk-7 Siberian Chemical Enterprise plutonium reprocessing facility, exploding a concrete bunker and exposing 160 onsite workers to excessive radiation	0	$44
January 14, 1995	Wiscasset, Maine, United States	Steam generator tubes unexpectedly crack at the Maine Yankee nuclear reactor, forcing the Maine Yankee Atomic Power Company to shut down the facility for 1 year	0	$62
February 2, 1995	Kota, Rajasthan, India	The Rajasthan Atomic Power Station leaks radioactive helium and heavy water into the Rana Pratap Sagar River, necessitating a 2-year shutdown for repairs	0	$280

(Continued)

Table 1: (*Continued*)

Date	Location	Description	Fatalities	Cost (in US$ million (2006))
May 16, 1995	Salem, New Jersey, United States	Ventilation systems fail at Public Service Electric & Gas Company's Salem Units 1 and 2	0	$34
February 20, 1996	Waterford, Connecticut, United States	Leaking valve forces the Northeast Utilities Company to shut down Millstone Units 1 and 2; further inspection reveals multiple equipment failures	0	$254
September 2, 1996	Crystal River, Florida, United States	Balance-of-plant equipment malfunction forces the Florida Power Corporation to shut down Crystal River Unit 3 and make extensive repairs	0	$384
September 5, 1996	Clinton, Illinois, United States	Reactor recirculation pump fails, prompting the Illinois Power Company to shut down the Clinton boiling water reactor	0	$38
September 20, 1996	Seneca, Illinois, United States	Service water system fails and prompts Commonwealth Edison to close LaSalle Units 1 and 2 for more than 2 years	0	$71
September 9, 1997	Bridgman, Michigan, United States	Ice condenser containment systems fail at Indiana Michigan Power Company's D.C. Cook Units 1 and 2	0	$11
May 25, 1999	Waterford, Connecticut, United States	Steam leak in feedwater heater causes manual shutdown and damage to the control board annunciator at the Millstone Nuclear Power Plant	0	$7
June 18, 1999	Shika, Ishikawa, Japan	Control rod malfunction sets off an uncontrolled nuclear reaction at Shika Nuclear Power Station's Unit 1	0	$34

(*Continued*)

Table 1: (*Continued*)

Date	Location	Description	Fatalities	Cost (in US$ million (2006))
September 29, 1999	Lower Alloways Creek, New Jersey, United States	Major freon leak at the Hope Creek Nuclear Facility causes the ventilation train chiller to trip, releasing toxic gas and damaging the cooling system	0	$2
September 30, 1999	Ibaraki Prefecture, Japan	Workers at the Tokaimura uranium processing facility try to save time by mixing uranium in buckets, killing 2 and injuring 1,200	2	$54
December 27, 1999	Blayais, France	An unexpectedly strong storm floods the Blayais-2 nuclear reactor, forcing an emergency shutdown after injection pumps and containment safety systems fail from water damage	0	$55
January 21, 2002	Manche, France	Control systems and safety valves fail after improper installation of condensers, forcing a 2-month shutdown	0	$102
February 16, 2002	Oak Harbor, Ohio, United States	Severe corrosion of control rod forces a 24-month outage of the Davis-Besse reactor	0	$143
October 22, 2002	Kalpakkam, India	Almost 100 kg of radioactive sodium at a fast breeder reactor leaks into a purification cabin, ruining a number of valves and operating systems	0	$30
January 15, 2003	Bridgman, Michigan, United States	A fault in the main transformer at the Donald C. Cook nuclear power plant causes a fire that damages the main generator and backup turbines	0	$10

(*Continued*)

Table 1: *(Continued)*

Date	Location	Description	Fatalities	Cost (in US$ million (2006))
April 10, 2003	Paks, Hungary	Damaged fuel rods hemorrhage spent fuel pellets, corroding a heavy water reactor	0	$37
August 9, 2004	Fukui Prefecture, Japan	Steam explosion at the Mihama Nuclear Power Plant kills 5 workers and injures dozens more	5	$9
April 19, 2005	Sellafield, United Kingdom	20 metric tons of uranium and 160 kg of plutonium leak from a cracked pipe at the Thorp nuclear fuel reprocessing plant	0	$65
May 16, 2005	Lorraine, France	Substandard electrical cables at the Cattenon-2 nuclear reactor cause a fire in an electricity funnel, damaging safety systems	0	$12
June 16, 2005	Braidwood, Illinois, United States	Exelon's Braidwood nuclear station leaks tritium and contaminates local water supplies	0	$41
August 4, 2005	Indian Point, New York, United States	Entergy's Indian Point Nuclear Plant, located on the Hudson River, leaks tritium and strontium into underground lakes from 1974 to 2005	0	$30
March 6, 2006	Erwin, Tennessee, United States	Nuclear fuel services plant spills 35 liters of highly enriched uranium, necessitating a 7-month shutdown	0	$98
December 24, 2006	Jadugoda, India	One of the pipes carrying radioactive waste from the Jadugoda uranium mill ruptures and distributes radioactive material more than 100 km^2	0	$25

(Continued)

Table 1: (*Continued*)

Date	Location	Description	Fatalities	Cost (in US$ million (2006))
July 18, 2007	Kashiwazaki, Japan	The Tokyo Electric Power Company announces that its Kariwa nuclear plant has leaked 1,192 liters of radioactive water into the Sea of Japan after being damaged by a 6.8-magnitude earthquake	0	$2
June 4, 2008	Ljubljana, Slovenia	Slovenian regulators shut down the Krsko nuclear power plant after the primary cooling system malfunctions and coolant spills into the reactor core	0	$1
June 14, 2008	Fukushima Province, Japan	A 7.2-magnitude earthquake cracks reactor cooling towers and spent fuel storage facilities, spilling 19 liters of radioactive wastewater and damaging the Tokyo Electric Power Company's No. 2 Kurihara Power Plant	0	$45
July 4, 2008	Ayrshire and Suffolk, United Kingdom	Two British Energy nuclear reactors (the Largs and the Sizewell B facilities) shut down unexpectedly after their cooling units simultaneously malfunction, damaging emergency systems and triggering blackouts	0	$10
July 13, 2008	Tricastin, France	The nuclear power operator Areva reports that dozens of liters of wastewater contaminated with uranium are being accidentally poured on the ground and run off into a nearby river	0	$7

(*Continued*)

Table 1: (*Continued*)

Date	Location	Description	Fatalities	Cost (in US$ million (2006))
March 15, 2009	Oskarshamn, Sweden	A maintenance worker repairing a shut-down reactor at the Oskarshamn Nuclear Power Plant dies after falling from the top of the turbine hall	1	$0
August 12, 2009	Gravelines, France	Assembly system fails to properly eject spent fuel rods from the Gravelines Nuclear Power Plant, causing the fuel rods to jam and the reactor to shut down	0	$2
August 27, 2009	St. Petersburg, Russia	A cracked discharge accumulator and malfunctioning feed pump force the Leningrad Nuclear Power Plant Reactor Number 3 to close for extended repairs	0	$110
February 1, 2010	Montpelier, Vermont, United States	Deteriorating underground pipes from the Vermont Yankee nuclear power plant leak radioactive tritium into groundwater supplies in Vermont, resulting in the eventual shutdown of the plant	0	$700

Note: An "accident" is defined as an incident that resulted in either the loss of human life or more than US$50,000 of property damage.

- A blown fuse on a sump pump at the Indian Point facility in New York, rendering the cooling system incapable of removing water leaking into the reactor, forcing a week-long shutdown;
- A technician testing for air leaks with a candle who accidentally dropped it and caused a fire that burned 1.6 million electrical cables, forcing a three-month shutdown at the Browns Ferry nuclear plant in Alabama;
- A nest of field mice causing an electrical fire that shut down the San Onofre nuclear facility in California for one week; and
- A three-year shutdown of the Davis-Besse nuclear power plant in Ohio after inspectors found excessive degradation of the pressure-vessel head of the reactor, but only after inspecting the wrong reactor by mistake.[25]

Other incidents include improper soldering preventing electricity and water from flowing properly in separate and supposedly independent backup systems; plastic floats that leaked, filled up, and sank (in that order) so that they all provided the same wrong indication of liquid level within the cooling system of a reactor; and supposedly independent equipment all being water-damaged from being stored together outdoors. Still others involve redundant safety systems all being disabled by the same contaminated lubricating oil, an entire system of independent cooling pipes all freezing because one thermostat on a protective heater had been improperly wired, and a four-week shutdown and a 24-hour blackout occurring after a stray cat wandered into a reactor vessel and shorted a circuit. The lesson appears to be that complicated technological systems, like reactors, have unavoidable problems — something Charles Perrow calls "normal accidents."[26]

Perrow argues that such accidents have three common themes, none of which bode well for nuclear power plants. First, no matter how well a system is designed, operator error is still a very common causal factor of system accidents. People are fallible, even those at nuclear power plants. Some nuclear employees will not be the best workers; many are fired for breaking the law, and in one case a construction crew put a safety inspector in the hospital for suggesting repairs. The Chair of the NRC once remarked that the nuclear industry "shows a surprising lack of professionalism."[27] The construction of one nuclear power plant in the US was prone to 111 separate flaws that cost an additional US$2.5 billion to repair; and employees and contractors used intimidation and deception, all part of industrial life, to try and hide their shoddy work.

Second, great events and accidents almost always have very small beginnings. Nuclear power plants are so complex that relatively simple things — shirttails, fuses, light bulbs, mice, cats, and candles — can disrupt the entire system. Even minor changes can have grave implications. Examples outside of the energy sector include a 29-cent switch that burned out by improper testing being responsible for the failure of Apollo 13, an O-ring causing the explosion of the *Challenger* in 1986, and a technician dropping a socket wrench down an Arkansas nuclear missile silo in 1980 and causing an explosion which ejected a Titan warhead into a nearby field. Given the right events, multiple and unexpected interactions can lead to system-wide failure.

Third, most failures are those of organizations more than technology, i.e. people managing or operating technologies in particular ways. Perhaps the best evidence here comes from a cross-cultural comparison of pressurized water reactors operating in the US, France, Germany, and Sweden, which found that cultural differences in operations have far-reaching implications for plant safety and nuclear power regulation.[28] The study identified meaningful differences in how facilities were regulated. The US style was more "command and control" and was based on strict adherence to the law, with inspectors forbidden to fraternize with staff, and with a highly legalistic, formalized system where operators rotated among many positions to be experts at every job. This contrasted with the European structure, which let operators themselves manage plants but used penalties and fines to punish wrongdoers, and also did not rotate people to different positions so they could become specialized in a particular area.

Within each nuclear facility itself, the study found distinct subcultures, generally distinguished by operators and engineers. The engineering subculture characterized a nuclear plant as an abstract, analytical, deterministic, and static system where events and sequences were linear. Engineers envisioned a nuclear plant as a predictable machine that could be broken down into its pieces. The operator subculture shared a physical, holistic, empirical, and dynamic view of the plant, envisioning it as more integrated. Operators viewed the facility as an organism, with feelings and even personalities and complex linkages that invariably deviated from predictions.

While not formally stated as a conclusion, the study did raise questions about nuclear safety. It noted that struggles frequently occurred between engineering and operator subcultures over authority, especially with respect to operating procedures and access to the control room. It found that every plant (and even every shift) had its own personality and its own set of cultures at work, making it difficult to standardize approaches to management. US plants operated more like submarines; German plants, like spaceships; and French plants, like sophisticated industrial machinery. Swedish plants were decorated with amusing trolls; German plants, with Blaupunkt radios; and French control rooms, with free coffee. The problem was that changes to operating procedures often failed to take into account these cultural differences, meaning they were in essence "tinkering with a real time experiment with possible irreversible consequences."[29] The study also noted that, over time, both operators and engineers developed a unique bond with the plant, and so they tended to discount problems in the face of their trust. At one German reactor visited, operators trusted the plant so much that, when alarms went off, they started diagnosing what was causing the alarms to malfunction instead of looking at the plant. In the US, operators believed they had the skill to pilot operations by hand, turning off automatic controls; and in France, operators treated alarms so casually that familiarity led to contempt for the consequences of errors.

Other key anthropological work on nuclear power safety culture has revealed a tendency to see incidents not as accidents, but rather as new sources of information about how a nuclear system functions. Perhaps perversely, some operators may even welcome and see value in accidents and incidents, as they contribute to new knowledge about reactor and human performance.[30] This can create the norm at a particular plant that a reactor is "safest when running," thus contributing to the pressure to keep plants online and delay maintenance, and also to consider production of electricity first and safety second. To these operators, nuclear work is supposed to be fast and efficient, not "slowed down" by frivolous safety concerns.[31]

All of the factors — operator error, the potential for failures to cascade, differences in organizational behavior, and the proclivity to ride margins to learn about nuclear performance — show the sensitivity of nuclear power plants to even slight deviations from normal activity. Perrow concludes that such high-tech, dangerous systems are hopeless and should be abandoned, as the inevitable risks of failure outweigh any conceivable benefits.[32] The point here is not that systems fail — no technology is perfect — but that nuclear power systems are so radioactive and catastrophic when they break down that a billion-dollar asset can become a trillion-dollar liability in a matter of moments. Such problems cannot be designed around. As more proof, consider that the two most significant nuclear power accidents, Chernobyl and Three Mile Island, were human-caused and then exacerbated by more human mistakes.

Chernobyl, Ukraine

On the evening of April 25, 1986, evening-shift engineers at Chernobyl's Reactor No. 4 experimented with the cooling pump system to see if it could still function without auxiliary electricity supplies. In order to proceed with the test, the operators turned off the automatic shutdown system. At the same time, they mistakenly lowered too many control rods into the reactor core, dropping plant output too quickly. This stressed the fuel pellets, causing ruptures and explosions, bursting the reactor roof and sweeping the eruption outwards into the surrounding atmosphere. As air raced into the shattered reactor, it ignited flammable carbon monoxide gas and created a radioactive fire that burned for nine days and continued to release radiation for more than two weeks.[33]

Following the accident, 116,000 people were evacuated from a 30-km^2 exclusive zone constituting parts of Belarus, Ukraine, and Russia. The large city of Prypiat, Ukraine, had to be completely abandoned (see Figure 1). The Chernobyl meltdown distributed more than 200 times the radiation released by the atom bombs dropped on Nagasaki and Hiroshima. More than 5 million people, including 1.6 million children, were exposed to dangerous levels of radiation. About 246,000 km^2 of land was contaminated with iodine-131, ruthenium-106, cerium-141 and cerium-144, cesium-137, strontium-89 and strontium-90, and plutonium-239 — some

Figure 1:   The City of Prypiat, Ukraine, Abandoned After the Chernobyl Disaster in 1986

of which will remain lethally radioactive for more than 10,000 years. At least 350,000 people had to be forcibly resettled from the area. Cesium and strontium severely contaminated agricultural products, livestock, and soil as far away as Japan and Norway; some milk in Eastern Europe is still undrinkable today.[34]

Human error after the initial accident also exacerbated the situation and needlessly exposed millions of people to unhealthy levels of radiation. For example, the Soviet government did not begin evacuations until April 28, two full days after the accident, because plant operators had delayed reporting the accident to Moscow out of fear it would spoil forthcoming May Day celebrations, and then because national officials had planned on covering it up until a Swedish radiation monitoring station 800 miles northwest of Chernobyl reported radiation levels 40% higher than normal. Russian and Ukrainian disaster managers mistakenly sent hundreds of buses contaminated with radioactive iodine during the evacuation back into public transportation service in Kiev. Some members of the Russian military personally contaminated themselves, and their families, by rushing back into the disaster area in what they believed was a sign of bravery. This act extended a long tradition of Soviet troops exposing themselves to radiation as a sign of strength, including tanks intentionally driving through sites of nuclear weapons fallout and aircraft flying back into the fallout from atmospheric weapons testing. In what could qualify as a scene from a National Lampoon movie if the consequences were not so dire, a Russian helicopter crew quickly redeployed from Afghanistan was assigned to drop boric acid on the exposed fissile material above Chernobyl's shattered reactor only to crash into it, causing yet another radioactive explosion.[35] The leader of the Soviet delegation charged with estimating damages (and also the one with the most expertise), Valery Legasov, later committed suicide on the second anniversary of the accident because he was so disturbed by his report's findings.[36]

After this accident (and the subsequent errors), traces of radioactive deposits unique to Chernobyl were found in nearly every country in the northern hemisphere. The international community sponsored a US$1.4 billion decontamination project, including the construction of a massive sarcophagus and 131 hydroelectric installations to prevent contaminated water from flowing downstream on the Prypiat and Dnieper rivers. Soviet authorities strongly urged as many as 400,000 abortions in an effort to mitigate the reporting of birth defects.[37] The International Atomic Energy Agency (IAEA), working with the World Health Organization (WHO), attributed up to 4,000 deaths to the Chernobyl nuclear accident; whereas other studies put the number at 93,000 fatal cancer deaths throughout Europe, 140,000 in Ukraine and Belarus, and another 60,000 in Russia, for a total of 293,000.[38] Medical studies have since confirmed that the thyroid cancer tumors caused by the Chernobyl accident are the largest single number of cancers of one type caused by a single event on one date ever recorded.[39] As the United Nations Scientific Committee on the Effects of Atomic Radiation concluded in 2000, the 25-fold increase in the incidence of childhood thyroid cancers in cities around Chernobyl showed that "there can be no doubt about the relationship between the radioactive materials released from the Chernobyl accident and the unusually high numbers of thyroid cancers observed in the contaminated areas during the past 14 years."[40]

The consequences of the accident at Chernobyl, moreover, are far from over. The fallout from Chernobyl contaminated about 6 million hectares of forest in the Gomel and Mogilev regions of Belarus, the Kiev region of Ukraine, and the Bryansk region of the Russian Federation.[41] Three of the contaminants — cesium-137, strontium-90, and plutonium-239 — are extraordinarily robust and extremely dangerous. About 95% of these contaminants have accumulated in living trees, but 770 wildfires occurred in the contaminated zone from 1993 to 2001, each one releasing radioactive emissions far into the atmosphere.[42] A single, severe fire in 1992 burned 5 km^2 of land contaminated by Chernobyl (including 2.7 km^2 of the highly contaminated Red Forest next to the reactor), carrying highly toxic cesium dust particles into the upper atmosphere, distributing radioactive smoke particles thousands of kilometers, and exposing at least 4.5 million people to dangerous levels of radiation. Radiation levels were so high after the 1992 fire that scientists throughout Europe initially thought there had been a second meltdown at Chernobyl Reactor No. 1 or 2, which remained in operation until 2000.

Three Mile Island

On March 28, 1979, equipment failures and operator error contributed to a partial core meltdown at the Three Mile Island (TMI) nuclear reactor in Pennsylvania, causing US$2.4 billion in property damages. Technically, the meltdown at TMI was a loss-of-coolant accident. The primary feedwater pumps stopped running at TMI Unit 2, preventing the large steam generators at the reactor site from removing necessary exhaust heat. As the steam turbines and the reactor automatically shut down, contaminated water poured out of open valves and caused the core of the reactor to overheat, inducing a partial core meltdown.[43]

A commission chartered by President Carter to study the accident, however, found that human error played the most significant factor in the meltdown.[44] The commission stated that the TMI operators were not well trained, operating procedures were confusing, and administrators had failed to learn lessons in safety from past incidents at the plant. The commission concluded that "we have stated that fundamental changes must occur in organizations, procedures, and above all, in the attitudes of people. No amount of technical 'fixes' will cure this underlying problem."[45]

Several American regulatory agencies have conducted detailed studies of the radiological consequences of this accident, and a consensus has emerged that — while the average dose of exposure from the accident was 1 millirem, or one-sixth the exposure from a full set of chest x-rays — the situation came dangerously close to releasing catastrophic amounts of radioactivity. For example, when federal investigators arrived on the scene, they discovered two pieces of alarming news that had not been widely reported. First, the reactor core was more badly damaged than previously thought. Falling coolant levels in the reactor core had exposed the tops of fuel rods to the air, causing oxidation of the cladding used to protect the rods. The result was that radioactive gases like xenon-133, krypton-85, and iodine-131 had seeped out of cracks in the reactor. Second, a gas bubble nearly 1,000 cubic feet in size had developed at the top of the reactor. Apparently, the reactor core had reached high enough temperatures that the coolant water had decomposed into its primary elements, hydrogen and oxygen. Investigators feared that the bubble would continue to grow, forcing even more coolant water out of the reactor and allowing the core to reach temperatures of 5,000°C; at that point, the uranium fuel would begin to melt, risking a total core meltdown and a catastrophic release of the reactor's radioactive material.[46] The fact that TMI could withstand such an incident was a fluke: it had a double containment shell capable of containing a hydrogen explosion only because the commercial flight path to Harrisburg Airport passed over the plant.[47]

Future Accidents

The incidents at Chernobyl and TMI brought about sweeping changes to the industry. After the accidents, emergency response planning, reactor operator training, human factors engineering, radiation protection, and many other areas of nuclear power plant operations were reformed. Yet despite these reforms, the risk of future accidents is still unacceptably high, and current operators appear to have forgotten some of these lessons.

For instance, the US GAO conducted a survey of nuclear power plant safety, since in the US 103 operating commercial nuclear power plants located at 65 sites in 31 states provide roughly 20% of the country's electricity. After conducting physical inspections of plant equipment and assessing indicators of plant performance, the GAO found that a number of individual nuclear power plants were not performing within acceptable safety guidelines.[48] Another sample of US plants inspected by the NRC from May 1999 to April 2004 revealed 25 serious incidents at 23 separate facilities.[49] Yet another study of US nuclear reactors identified nearly 60 accidents or near misses — events that included radiation exposure, inhalation of toxic vapors, electrical shocks, and injuries during nuclear construction or maintenance — resulting "in serious worker injuries or facility damage."[50] Still another study performed by the agency in charge of monitoring the US Department of Energy's oversight of nuclear facilities, including research reactors and national laboratories, found that 31 of the 205 facilities did not meet government safety requirements, and that one-third of the facilities did not conform to guidelines concerning high-hazard nuclear waste.[51] The GAO identified 156 serious incidents from 2001 to 2005 at US nuclear power plants that included a litany of problems, ranging from unplanned changes in reactor power and failures of emergency diesel generators to inadequate maintenance and human mismanagement.[52]

Most recently, a 2009 assessment of nuclear power performance in the US warned that the technology has rushed "far ahead of its operating experience." The study noted that every state across the northern tier from Illinois to Maine has been involved in at least one nuclear accident. These involved quality assurance breakdowns, plant equipment sinking into the mud, fuel cladding failures, emergency core cooling system shortcomings, absence of emergency plans, radioactive leaks, and water contamination.[53] Safety problems have not been recognized; when they have been recognized, they have not been resolved; and the industry has not made significant strides towards addressing newly emerging threats like terrorism. As two environmental lawyers recently put it, "the nuclear industry in the United States is like the financial industry was prior to the crisis of 2008; there are many risks that are not being properly managed or regulated."[54]

These safety problems take their toll on performance and contribute to the likelihood of future accidents. Using some of the most advanced probabilistic risk assessment tools available, an interdisciplinary team at the Massachusetts Institute of Technology (MIT) identified possible reactor failures in the US and predicted that the best estimate of core damage frequency was around one every 10,000 reactor years. In terms of the expected growth scenario for nuclear power from 2005 to 2055, the MIT team estimated that at least four serious core damage accidents will occur and concluded that "both the historical and the PRA [probabilistic risk assessment] data show an unacceptable accident frequency." Furthermore, "[t]he potential impact on the public from safety or waste management failure ... make it impossible today to make a credible case for the immediate expanded use of nuclear power."[55]

Another assessment conducted by the CEA (*Commissariat à l'énergie atomique et aux énergies alternatives*) in France tried to associate nuclear plant design with human error such that technical innovation could help eliminate the risk of human-induced accidents.[56] Two types of mistakes were deemed the most egregious: errors committed during field operations (such as maintenance and testing) that can cause an accident, and human errors made during small accidents that cascade to complete failure. There may be no feasible way to "design around" these risks. For example, when CEA researchers examined the safety performance of advanced French pressurized water reactors, they concluded that human factors would contribute to about one-fourth (23%) of the likelihood of a major accident.

Because of the lack of a permanent geologic repository for nuclear waste in the US at offsite storage facilities, operators have packed spent fuel more densely together at existing onsite storage pools. In some cases, such densities even approach those found in operating reactor cores and, if exposed to air for more than six hours, spent fuel rods will combust spontaneously.[57] Robert Alvarez and his colleagues have warned that recently discharged fuel could heat up rapidly and cause a fuel cladding fire that would disperse volatile fission products, such as cesium-137, over hundreds of miles.[58] Cooling water at these pools could be lost — the precursor to a fire — in a variety of ways: draining into other volumes through a malfunctioning of valves, pipes and gates breaking and unable to hold water, a large aircraft crash puncturing a shaft to cause leakage, or a shaped charge of explosive cracking concrete vessels.

Oddly, there is some evidence that the *newest* reactors and nuclear systems are the most prone to accidents. Dennis Berry, Director Emeritus of

Sandia National Laboratories, explained that the problem with new reactors and accidents is twofold: scenarios arise that are impossible to plan for in simulations, and humans make mistakes. As he put it, "fabrication, construction, operation, and maintenance of new reactors will face a steep learning curve: advanced technologies will have a heightened risk of accidents and mistakes. The technology may be proven, but people are not."[59]

Indeed, nuclear engineer David Lochbaum noted that almost all serious nuclear accidents have occurred when operators have little experience with a plant, in essence making newest systems the riskiest.[60] In 1959, the Sodium Research Experiment reactor in California experienced a partial meltdown 14 months after opening. In 1961, the Sl1 reactor in Idaho was slightly more than two years old before a fatal accident killed everyone at the site. The Fermi Unit 1 reactor began commercial operation in August 1966, but had a partial meltdown only two months after opening. The St. Laurent des Eaux A1 reactor in France started in June 1969, but an online refueling machine malfunctioned and melted 400 pounds of fuel four months later. The Browns Ferry Unit 1 reactor in Alabama began commercial operation in August 1974, but experienced a fire that severely damaged control equipment six months later. The TMI Unit 2 reactor began commercial operation in December 1978, but had a partial meltdown three months after it started. The Chernobyl Unit 4 reactor started up in August 1984, but suffered the worst nuclear disaster in history on April 26, 1986, before the two-year anniversary of its operation. NRC files show about 3,000 incidents and events involving abnormal occurrences and violations of safety regulations each year at American nuclear plants, with most of these peaking in the 1970s (when the commercial nuclear fleet was the youngest).[61] The implication is that when nuclear power plant designs are "new," errors are more likely, not less.

Outside of the US, new designs such as the Superphénix fast breeder reactor in France — considered to be the cutting edge of nuclear technology — have been plagued by breakdowns, disappearing fuel, and other assorted problems so severe that the reactor operated only 174 days out of its first eight years.[62] The breeder reactor at Marcoule, France, was shut down in 1997 after a series of fuel rods jammed and could not be removed. Liquid sodium being emptied from the Rhapsodie breeder reactor caused an explosion that lifted up concrete slabs and sent them into the air, where they crashed into and killed operators.[63] The British similarly abandoned their prototype fast reactor after safety problems, and the Russians stopped research on their BN-350 reactor over design flaws. A German breeder reactor at Kalkar was completed in 1991, but never operated because of concerns about explosions; and the Japanese Monju reactor was shut down after a serious fire in December 1995.[64] India's fast breeder test reactor at Kalpakkam has seen multiple pump failures, shutdowns due to faulty instrument signals, and turbine malfunctions. In May 1987, two years after it started operation, an accident occurred during refueling whereby 23 fuel assemblies were knocked out of the core, causing a two-year shutdown.[65] Another accident occurred in 2002 after a defective valve leaked 75 kg of radioactive sodium into a purification cabin.[66] The safety assessment from the Atomic Energy Regulatory Board of India has long considered it the most dangerous reactor in the country.[67]

Another advanced design, the European pressurized water reactor, called the "flagship" of the next generation of designs, has encountered similar problems in Olkiluoto, Finland, an island on the Baltic Sea, where it is still being built. The Finnish nuclear safety watchdog STUK has reported 2,100 quality defects in the plant so far, and the project is US$2.4 billion overbudget and three years behind schedule.[68] Serious problems have arisen over the vast concrete foundation of the reactor building, which was found too porous and prone to corrosion.[69] Although the reactor was originally meant to be completed in the summer of 2009, Areva, the French company building it, will no longer make predictions on when it will be finished. In Flamanville, France, a "clone" of the Finnish reactor now under construction is also behind schedule and overbudget. There, nuclear safety inspectors discovered cracks in the concrete base and steel reinforcements installed in the wrong areas, and warned Électricité de France that the welders working on reactor components were not properly qualified.[70]

Indeed, several other factors seem to increase the risk of future accidents. The pressure to build new generators on existing sites to avoid complex issues associated with finding new locations can increase the risk of catastrophe, since there is a greater chance that one accident can affect multiple reactors. Nuclear power plants used to be sited more remotely (meaning an accident would affect less people), but now tend to be sited closer to population centers.[71] Sites that had once been remote when reactors started have become more populated over time; regions in which nuclear power is most attractive tend to be urban and have a limited number of remote locations; and substantial losses and costs are associated with remote transmission of nuclear power, creating an incentive to situate plants closer to points of electricity consumption. Nuclear waste storage is also becoming more dangerous, with many spent fuel pools packed with more fuel rods, making them hotter and more dense — hence, operators must add boron to the water pools to absorb neutrons, increasing the risk of criticality accidents.[72]

In addition, the industry has been trying to scale up reactor sizes and promote designs that operators have little experience with. These larger reactors tend to use more fuel and create more heat, meaning they have bigger cores containing larger quantities of dangerous fissionable materials, increasing the magnitude of any accident that could occur. The restructuring of electricity sectors around the world has motivated some nuclear operators to place profits before safety. Undue solicitude for profits of the licensee has played a large role in explaining the mishaps that have occurred at nuclear power plants. Put another way, nuclear power is least safe in environments where complacency and pressure to maximize profits are the greatest, yet the global trend appears headed in that direction.[73] It appears one can build nuclear power plants to be safe or to be cheap, but not both.

Lastly, some operators have begun to promote the automation of certain tasks in the control room to decrease human workload and improve system performance. However, numerous examples exist whereby automotive systems have only increased incidents and accidents, reduced operator awareness, increased monitoring and workload, and degraded manual operating skills.[74]

These factors are worrying, to say the least, given the outrageous severity of what a single serious accident can do. It was estimated that the meltdown of a single 500-MW reactor located 30 miles from a single city would cause the immediate death of between 3,400 and 45,000 people, injure roughly another 50,000, induce US$7 billion in property damage, and contaminate an area the size of Pennsylvania with unsafe levels of radiation.[75] This estimate was later revised upwards for being too conservative to 45,000 immediate deaths, 70,000 injuries, and US$17 billion or more in damages. More recent studies, looking at larger reactors in the 1,100-MW range, have noted that as many as 103,000 immediate fatalities could occur along with US$300 billion in damages.[76] A successful attack on the Indian Point power plant near New York City, apparently part of Al-Qaeda's original plan on September 11, 2001, would have resulted in 43,700 immediate fatalities and 518,000 cancer deaths and would have cost US$2 trillion to clean up.[77]

To put a serious accident in context, if 10 million people were exposed to radiation from a nuclear meltdown, about 100,000 would die from acute radiation sickness in six weeks. About 50,000 would experience acute breathlessness and 240,000 would develop acute hypothyroidism. About 350,000 men would be temporarily sterile, 100,000 women would stop menstruating, 100,000 children would be born mentally retarded, and there would be thousands of spontaneous abortions and more than 300,000 cancers to develop later.[78] In the US, the impact of an accident could be quite acute, since 80 million Americans live within 40 miles of a nuclear reactor, including those residing in some of the largest metropolitan areas in the country such as New York, Chicago, Detroit, Miami, Phoenix, Cleveland, Houston, and Philadelphia.[79]

A nuclear meltdown or reactor accident is not the only thing to worry about. The long-term impacts of a fuel cladding fire could be significantly *worse* than those of Chernobyl, with hundreds of billions of dollars of damage in addition to human deaths and environmental damage.[80] Storage facilities are not located inside containment buildings and are usually aboveground, making the consequences of an accident more severe. As one peer-reviewed study put it: "A loss of coolant ... [at a storage pool] could result in a nuclear accident. A loss of coolant could cause rapid heating, and then the outside shell of the fuel rods could catch fire. The result could be significant dispersal of highly radioactive fission products."[81] Even the NRC estimated that the median consequences of a spent-fuel fire at a pressurized water reactor would result in 54,000–143,000 extra cancer deaths, 2,000–7,000 km^2 of agricultural land contaminated, and economic costs due to evacuation as high as US$566 billion.[82] Another study projected that one single pool fire would cause 24,000 lung cancer deaths and induce economic damages ten times as large as those caused by Hurricane Katrina.[83]

Materials and Labor

Separate from the risk of incidents and accidents, the lack of qualified and experienced nuclear staff serves as a technical challenge facing the nuclear industry. As Chapter 2 noted, a typical nuclear power plant can take 5–10 years to build; contain miles of pipes with thousands of welds; and require almost 1,000 miles of electrical cables and a prodigious number of electric motors, fuses, and circuits as well as radiation shields, spent fuel repositories, backup electricity generators, and firewalls. Building a single plant takes an enormous amount of expertise and about 10,000 dedicated construction workers, and the existing nuclear industry already lacks qualified and experienced staff. The global nuclear industry continues to lose much of the expertise that it does have to retirement, attrition, and death.

The Nuclear Energy Agency surveyed 16 nuclear member states of the Organisation for Economic Co-operation and Development (OECD), and concluded that some countries were "at risk" due to lack of educational capability for training in nuclear-related fields.[84] It documented declining university enrollment in nuclear engineering courses, an overall aging of the nuclear workforce, dilution of university course content related to nuclear physics, and changing expectations among young engineers that predisposed them away from working at nuclear power plants. As the study noted, "the nuclear industry does not attract the high numbers of good quality graduates and post graduates as it did when it was a fast developing and emerging industry a number of years ago. There are also problems with the retention of younger trained staff who are readily marketable to other sectors after a period in the nuclear industry."[85]

In the US, the Department of Energy (DOE) has warned that the lack of growth in the domestic nuclear industry has gradually eroded important infrastructural elements such as experienced personnel in nuclear energy operations, engineering, radiation protection, and other professional disciplines; qualified suppliers of nuclear equipment and components, including fabrication capability; and contractor, architect, and engineer organizations with personnel, skills, and experience in nuclear design, engineering, and construction.[86] Since all commercial American reactors are light water reactors, system operators have little experience with newer gas-cooled and other advanced reactor designs used throughout the world. Only two companies in the world, Japan Steel Works and Creusot Forge, currently have the heavy forging capability to create the largest reactor components.[87] In the 1970s, more than 400 suppliers of nuclear plant components existed, but the number dropped to 80 suppliers in 2008. Moreover, the Nuclear Energy Institute cautioned in 2005 that "half of the industry's employees are over 47 years old, and more than a quarter ... already are eligible to stop working," implying that the industry has far fewer available specialists with the requisite knowledge necessary to facilitate any rapid expansion of nuclear power.[88]

Another assessment in the UK warned that "the nuclear industry is facing a skills crisis."[89] As of 2007, fewer than 6% of the estimated 100,000 people who worked in the industry were under the age of 24; and at British Energy, which operates eight nuclear power stations and is the country's biggest electricity provider, 40% of the staff are set to retire within the next ten years. No British university offered a dedicated nuclear engineering course as of 2007 and a number of "vital occupations" remained unfilled. One nuclear consultant found it "amazing that so many people jumped on the bandwagon of this renaissance without ever looking at the industrial side of it." Another industry survey identified an "under-supply of qualified people" compared to the proportion of needed jobs in the nuclear power sector. It found a 20% deficit for high-level jobs, especially related to decommissioning, process and machine operators, and senior managers. The National Skills Academy for Nuclear in the UK estimated that as many as 16,500 new workers would be needed to operate existing facilities by 2015. The survey warned that "the sector needs to quadruple the number of apprentices over the next five years."[90]

Given these constraints in human capital, the fastest deployment of nuclear reactors for a single country has been France. France, which currently generates about three-quarters of its electricity from nuclear units, has the quickest record for deploying nuclear plants in history: 58 plants between 1977 and 1993, or an average of 3.4 reactors per year. The fastest the US ever deployed nuclear power plants was a peak in 1974 at 12 per year, and the greatest number of nuclear power plants being built at once globally was 28 in 1984. Yet to meet the target by 2030, slightly more than 1,900 new nuclear plants (sized 1,000 MW each) would need to be built — or a minimum of 86 plants per year every year for more than two decades, greater than three times the fastest historical rate on record.

Fuel Availability and Energy Payback

As a third and final technical impediment, almost all commercial nuclear reactors (even those that utilize reprocessed fuel) need fresh uranium ore to operate, something that lowers their overall energy payback — the amount of net energy produced from the overall nuclear fuel cycle. The IAEA classifies uranium broadly into two categories: "primary supply," including all newly mined and processed uranium; and "secondary supply," encompassing uranium from reprocessing inventories, including highly enriched uranium, enriched uranium inventories, mixed oxide fuel, reprocessed uranium, and depleted uranium tails. The IAEA, after collecting information on 582 uranium mines and deposits worldwide, expected primary supply to cover 42% of the demand for uranium in 2000, but acknowledged that the number will drop to between 4% and 6% of supply in 2025, as low-cost ores are expended and countries are forced to explore harder-to-reach, more expensive sites.[91] However, here lies a conundrum: the IAEA calculated that secondary supply can only contribute 8–11% of world demand. "As we look to the future, presently known resources fall short of demand," the IAEA stated in 2001, and "it will become necessary to rely on very high cost conventional or unconventional resources to meet demand as the lower cost known resources are exhausted."[92]

There simply will not be enough uranium to go around, even under current demand. Interestingly, though, the IAEA refused to state this obvious conclusion. While the IAEA recorded the total amount of uranium at around 3.6 Gg in 2001, the number inexplicably jumped to 4.7 Gg in 2006. The increase was not due to new discoveries or improved technologies, but rather because of a clever redefinition of what the IAEA counted as uranium. The IAEA included in its new estimate the category of uranium that costs US$80–$130 per kilogram. This class comprises uranium ores that are of relatively low grade and of greater depth so much harder to mine, and that require such longer transport that the IAEA historically has not even counted them as usable stocks of uranium at all.

Another October 2008 assessment reported that the world presently consumes 160 million pounds of uranium per year to fuel existing reactors, but only produces 100 million pounds.[93] The difference is made up from stored inventories of mined uranium, unused fuel from decommissioned plants, and diluted nuclear weapons, but these reserves are being exhausted.

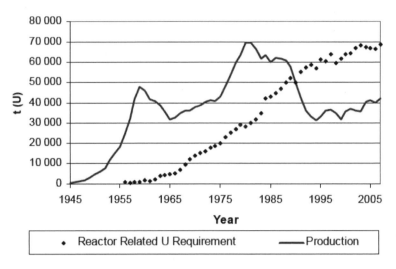

Figure 2: Global Annual Uranium Production and Reactor Fuel Requirements (in tons), 1945–2005[94]

As one example, the US produced 4.5 million pounds of uranium in 2007 but had to import 47 million pounds (or ten times as much) from other countries. The assessment concluded that enough high-grade uranium ore exists to supply the needs of the current fleet for 40–50 years, but warned that if the construction of new nuclear power plants were to accelerate, existing resources would not last more than ten years. Figure 2 clearly shows that a mismatch between uranium supply and demand is thus emerging, with dependence on secondary sources of uranium believed to run out by 2015.

One study from the Institute of Particle Physics of ETH Zurich and CERN (European Organization for Nuclear Research) cautioned that extraction from known mines and secondary resources during the coming 5–10 years appears to be much more difficult than generally believed, and almost no country that uses nuclear energy is self-sufficient in fuel production. Table 2, for example, shows that virtually every country producing uranium is now past its peak.[95] Germany and France have essentially stopped uranium mining; Japan, the UK, South Korea, and Sweden never had any substantial mining operations of their own; and production in the US is not even sufficient to satisfy 10% of national demand.

Table 2:   Global Uranium Production and Use for Nuclear Reactors, 2008

Country	Nuclear Power Capacity (GWe)	Total Uranium Produced	Total Uranium Required	Peak Production (tons) and Year
United States	99	1,430	18,918	16,811 (1980/1981)
France	63.5	5	10,527	3,394 (1987/1988)
Japan	47.6	0	7,569	10 (1972/1973)
Russian Federation	21.7	3,521	3,365	16,000 (1987/1988)
Germany	20.3	0	3,332	7,090 (1965/1966)
South Korea	17.5	0	3,109	—
Ukraine	13.1	800	1,974	1,000 (1992/1993)
Canada	12.6	9,000	1,665	12,522 (2001/2002)
United Kingdom	11	0	2,199	—
Sweden	9	0	1,418	29 (1969)
South Africa/Namibia	1.8	5,021	303	10,188 (1980/1981)
Australia	0	8,430	0	9,512 (2004/2005)
Kazakhstan	0	8,521	0	8,521 (2008/2009)
Niger	0	3,032	0	4,363 (1981/1982)
World	**372**	**43,853**	**65,000**	**69,692 (1980/1981)**

Some Asian countries, such as China and India, have domestically available supplies of uranium, but these are extremely limited. The China National Nuclear Corporation expects the country's demand for uranium to rise from 1,000 tons per year in 2007 to 7,000 tons by 2020, by which time China will be more dependent on Australia for uranium imports. In fact, Chinese officials have already signed a deal with Australian firms to import 20,000 tons of uranium by 2020. Supplies of uranium ore are now recognized as "probably the biggest hurdle to expansion of the mainland's nuclear sector," and Chinese analysts expect the country to be dependent on foreign sources for 88% of its uranium ore by 2020.[96]

Geologists have estimated that India has about 61,000 tons of uranium reserves, but caution that most of it is stranded — far from existing mines and reactors where fuel is needed — and of very poor quality. Uranium mining companies have argued that Indian uranium ore concentrations hover around the 0.06% mark, compared to the minimum "economically exploitable" concentration of 0.1%. This dearth of recoverable Indian uranium has convinced many engineers to talk about shifting to thorium fuel cycles, but such advanced technology is at least a few decades away. Moreover, domestic Indian uranium supplies are already insufficient to supply existing nuclear power plants. Operators shut down five of the 17 nuclear power plants in the country at the end of 2007 and operated the remaining reactors at less than 50% capacity for want of fuel. Uranium fuel shortages have also forced the Nuclear Power Corporation of India to delay the commissioning of two new units at the Rajasthan Atomic Power Station and another new unit at Kaiga in Karnataka.[97]

To summarize, for the past 15 years only about two-thirds of global uranium requirements (between 31,000 and 44,000 tons) have been extracted from actual uranium mines, with the shortfall made up of civilian and military stocks of uranium and plutonium built up during the Cold War along with mixed oxide reprocessing. These secondary sources, however, are becoming rapidly exhausted, thus convincing the Nuclear Energy Agency and the IAEA to declare that "most secondary resources [of uranium] are now in decline and the gap will increasingly need to be closed by new production. Given the long lead time typically required to bring new resources into production, uranium supply shortfalls could develop."[98]

Even the reserves from existing mines are being rapidly depleted. One assessment from the IAEA, hardly an organization against nuclear power, concluded that enough high-grade uranium ore exists to supply the needs of the current fleet for only 40–50 years and warned that, if the construction of new nuclear power plants were to accelerate so that all coal plants were replaced, existing resources would not last more than ten years.[99] The most recent assessment of uranium deposits published by the IAEA noted that reasonably assured resources of uranium amounted to less than 3.4 million tons of uranium in 2009 (see Table 3) — enough to supply the existing reactor fleet for only 83 years, assuming annual production remained constant at 40,260 tons.[100] The US DOE has quietly acknowledged that domestic uranium production is currently at about 10% of its historical peak, and that most of the world's uranium reserves are becoming "stranded" and therefore much more difficult to extract.[101]

Such a bleak outlook was recently confirmed by a peer-reviewed study on available uranium resources at 93 deposits and fields located in Argentina, Australia, Brazil, Canada, Central African Republic, France, Kazakhstan, Malawi, Mongolia, Namibia, Niger, Russia, South Africa, the US, and

Table 3:   Reasonably Assured Resources of Uranium by Country
(tons of uranium below US$130/kg U)

Country	Reasonably Assured Resources	%
Australia	725,000	22
Kazakhstan	378,000	11
United States	339,000	10
Canada	329,200	10
South Africa	284,400	8
Niger	243,100	7
Namibia	176,400	5
Russia	172,400	5
Brazil	157,400	5
Uzbekistan	72,400	2
India	48,900	1
China	48,800	1
Others	363,300	13
Total	3,338,300	100

Zambia.[102] The study reported that the quality of mined uranium peaked during the nuclear weapons programs of the 1940s and 1950s, when the highest-grade deposits were depleted. A long-term decline in the average uranium ore grade for almost all suppliers was documented. In the US, for example, the quality of uranium dropped from an average of 0.28% U_3O_8 in 1980 to 0.09% in 2005 — a decline of one third despite improvements in technology. No "world class" discoveries of uranium have occurred since the 1980s, and all increases in uranium mining and milling between 1988 and 2005 resulted from increased drilling and new assessments at known deposits. The study noted that uranium miners are having to go deeper and use more energy and water to extract uranium resources as the overall quality of ore declines.

To further complicate matters, finding and developing new deposits and fields requires large amounts of time and significant capital, as new mines and enrichment facilities take longer than a decade to bring online and can be delayed (like nuclear reactors) by unforeseen events. For example, a single cyclone stopped production of the Australian Ranger open-pit uranium mine in 2007 for more than a year, and the completion of the

Cigar Lake uranium mine in Canada was delayed for three years by a flash flood.[103]

Researchers at the Oxford Research Group suggest that declining ore grades will eventually yield a negative net energy loss before the end of this century.[104] They posit that the energy required to enrich ores of less than 0.02% U_3O_8 exceeds the total energy the uranium can produce. The global average for ore grade currently stands at 0.15%, though the range varies tremendously from high-grade locations like Canada's McArthur River (>21%)[105] to Australia's Lake Maitland (0.04%).[106] Although reserves today are comfortably above the 0.02% threshold, as high-grade reserves are exhausted, production will shift to low-grade sources at a higher cost. While technological advances may enable profitable access to these resources, doing so will inevitably need more energy and thus a larger carbon footprint.

The declining availability of high-quality uranium fuel, along with other factors, contributes to nuclear energy having a low energy payback. Even utilizing the richest ores available, a nuclear power plant must operate at ten full-load operating years before it has paid off its energy debts.[107] Based on this estimation, several known facts can modify the calculation: not all plants use the richest ores, plants operate at full capacity for an average of only 20 years, and most plants are decommissioned within 30 or 40 years. Accordingly, a plant using average-quality uranium and operating at full capacity for 20 years out of a 35-year life span will only generate twice as much energy as that consumed by the plant.

Other studies have documented how nuclear power plants generate 16% of global electricity, but provide only 6.3% of energy production and 2.6% of final energy consumption.[108] What accounts for this mismatch between generation, production, and consumption? Part of it stems from the poor consumption efficiency of electricity compared to other energy carriers as a whole (as electricity is relatively inefficient in energy terms compared to the use of oil). The other part relates to transmission losses associated with nuclear energy, usually situated far away from sources of demand, as well as the energy used by nuclear plants themselves (for cooling, management of spent fuel, operations, and refueling).

Utilizing a similar technique called an "energy payback ratio" (i.e. the ratio of total energy produced compared to the energy needed to build, maintain, operate, and fuel an energy system), Luc Gagnon found that nuclear power plants score unfavorably. He estimated that hydroelectric, wind, and biomass power plants are at least 1.5–20 times more efficient from an energy payback perspective than nuclear reactors.[109] Another meta-survey of hundreds of energy payback ratio studies found that hydroelectric facilities had the best performance (with ratios exceeding 170), and that biomass and wind power plants performed well (27–34) compared to ratios of below 16 for nuclear power plants and below 7 for fossil-fueled plants.[110] Figure 3 shows the energy payback ratios for a broad spectrum of technologies. Why do nuclear and fossil-fueled systems have such low energy payback ratios? As the best oil, gas, and uranium reserves get depleted, they tend to be replaced by wells and mines that require a higher energy investment (located in faraway regions). This leads to longer delivery distances and more energy needed for distribution. Other estimates have also confirmed nuclear's poor energy payback ratio compared to renewables such as wind and hydro.[111]

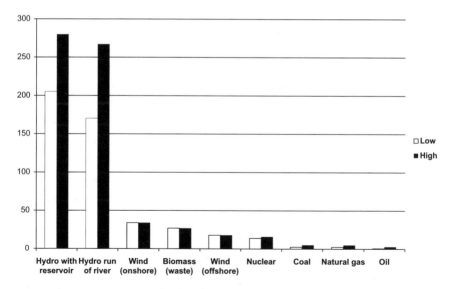

Figure 3:   Energy Payback Ratios for Various Technologies and Systems

Note: A high ratio indicates good environmental performance. If a system has a payback ratio between 1 and 1.5, it consumes nearly as much energy as it generates.

These technical challenges alone — scores of incidents and accidents, a high probability of future accidents, reactor meltdowns, fuel cladding fires, an aging workforce and lack of skilled staff, and a declining energy payback ratio and uncertain reserves of fuel — might be sufficient to stop a nuclear renaissance on their own. Yet as the pages to follow will show, a nuclear renaissance must also overcome immense economic, environmental, and sociopolitical hurdles if it is to become a reality.

Endnotes

[1] Peter A. Bradford, "Three Mile Island: Thirty Years of Lessons Learned," Testimony before the Senate Committee on Environment and Public Works (March 24, 2009).

[2] *Ibid.*

[3] Denis E. Beller, "Atomic Time Machines: Back to the Nuclear Future," *Journal of Land, Resources, & Environmental Law* 24 (2004), p. 43.

[4] Phillip A. Greenberg, "Safety, Accidents, and Public Acceptance," in John Byrne and Steven M. Hoffman (eds.), *Governing the Atom: The Politics of Risk* (London: Transaction Publishers, 1996), pp. 127–175.

[5] International Institute for Strategic Studies, *Preventing Nuclear Dangers in Southeast Asia and Australasia* (London: IISS, September 2009).

[6] See Stefan Hirschberg, Gerard Spiekerman, and Roberto Dones, *Severe Accidents in the Energy Sector* (1st edition) (Villigen, Switzerland: Paul Scherrer Institute, November 1998, PSI Report No. 98-16); Stefan Hirschberg and Andrej Strupczewski, "Comparison of Accident Risks in Different Energy Systems: How Acceptable?," *IAEA Bulletin* 41 (January, 1999), pp. 25–30; and Stefan Hirschberg, Peter Burgherr, Gerard Spiekerman, and Roberto Dones, "Severe Accidents in the Energy Sector: Comparative Perspective," *Journal of Hazardous Materials* 111 (2004), pp. 57–65.

[7] Benjamin K. Sovacool, "The Costs of Failure: A Preliminary Assessment of Major Energy Accidents, 1907–2007," *Energy Policy* 36 (2008), p. 1807.

[8] Christopher P. Winter, "Accidents Involving Nuclear Energy," available at http://www.chriswinter.com/Digressions/Nuke-Goofs/ (last visited November 6, 2008).

[9] Zachary Smith, *The Environmental Policy Paradox* (Upper Saddle River: Prentice Hall, 2009).

10 American University, *TED Case Studies: Environmental Threats of Russian Nuclear Trade* (Washington, D.C.: American University, 1996, Case No. 342), available at http://www.american.edu/TED/russnuke.htm/ (downloaded March 10, 2009).

11 Greenpeace, *France's Nuclear Failures: The Great Illusion of Nuclear Energy* (Amsterdam: Greenpeace International, November 2008), p. 10.

12 US Atomic Energy Commission of Occupational Safety, *Operational Accidents and Radiation Exposure Experience Within the United States Atomic Energy Commission, 1943–1975* (US Atomic Energy Commission, January 1, 1975, WASH-1192).

13 Constance Perin, "Operating as Experimenting: Synthesizing Engineering and Scientific Values in Nuclear Power Production," *Science, Technology, & Human Values* 23(1) (Winter, 1998), pp. 98–128.

14 Michele Adato *et al.*, *Safety Second: The Nuclear Regulatory Commission and America's Nuclear Power Plants* (Bloomington, IN: Indiana University Press, 1987).

15 Hirschberg and Strupczewski (1999), pp. 25–31.

16 Amory B. Lovins and L. Hunter Lovins, *Brittle Power: Energy Strategy for National Security* (Andover, MA: Brick House, 1982), pp. 157–158.

17 Barbara Rose Johnston, Susan E. Dawson, and Gary E. Madsen, "Uranium Mining and Milling: Navajo Experiences in the American Southwest," in Laura Nader (ed.), *The Energy Reader* (London: Wiley-Blackwell, 2010), pp. 132–146.

18 International Institute for Strategic Studies (2009).

19 Allison Macfarlane, "Interim Storage of Spent Fuel in the United States," *Annual Review of Energy and Environment* 26 (2001), pp. 201–235.

20 Public Citizen, *The Big Blackout and Amnesia in Congress: Lawmakers Turn a Blind Eye to the Danger of Nuclear Power and the Failure of Electricity Deregulation* (2003), p. 4.

21 Mohammad Saleem Zafar, *Vulnerability of Research Reactors to Attack* (Washington, D.C.: Stimson Center, April 2008).

22 US Government Accountability Office, *Nuclear Safety: Department of Energy Needs to Strengthen Its Independent Oversight of Nuclear Facilities and Operations* (Washington, D.C.: US GAO, October 2008, GAO 09-61).

23 *Ibid.*, p. 3.

24 Data taken from Benjamin K. Sovacool, "A Critical Evaluation of Nuclear Power and Renewable Energy in Asia," *Journal of Contemporary Asia* 40(3) (August, 2010), pp. 369–400.

25 Charles Perrow, *Normal Accidents: Living with High-Risk Technologies* (New York: Basic Books, 1984); and International Institute for Strategic Studies (2009), p. 36.

26 Perrow (1984).

27 *Ibid.*, p. 37.

28 Gene I. Rochlin and Alexandra von Meier, "Nuclear Power Operations: A Cross-Cultural Perspective," *Annual Review of Energy and the Environment* 19 (1994), pp. 153–187.

29 *Ibid.*, p. 185.

30 Perin (1998).

31 *Ibid.*

32 Perrow (1984), p. 304.

33 David R. Marples, "Nuclear Politics in Soviet and Post-Soviet Europe," in John Byrne and Steven M. Hoffman (eds.), *Governing the Atom: The Politics of Risk* (London: Transaction Publishers, 1996), pp. 247–270.

34 Benjamin K. Sovacool and Christopher Cooper, "Nuclear Nonsense: Why Nuclear Power Is No Answer to Climate Change and the World's Post-Kyoto Energy Challenges," *William & Mary Environmental Law and Policy Review* 33(1) (Fall, 2008), pp. 1–119.

35 *Ibid.*

36 Marples (1996), p. 255.

37 Douglas Chapin, Karl Cohen, W.K. Davis, E. Kinter, L. Loch *et al.*, "Nuclear Power Plants and Their Fuel as Terrorist Targets," *Science* 297 (September 20, 2002), pp. 1997–1999.

38 *Ibid.*

39 David J. Brenner, "Revisiting Nuclear Power Plant Safety," *Science* 299 (January 10, 2003), pp. 201–203.

40 UN Scientific Committee on the Effects of Atomic Radiation (UNSCEAR), *Sources and Effects of Ionizing Radiation: UNSCEAR 2000 Report to the General Assembly* (New York: United Nations, 2000).

41 Sergey I. Dusha-Gudym, "Transport of Radioactive Materials by Wildland Fires in the Chernobyl Accident Zone: How to Address the Problem," *International Forest Fire News* (January–June, 2005), p. 119.

42 Ryszard Szczygiel and Barbara Ubysz, "Chernobyl Forests, Two Decades After the Contamination," *Przeglad Pozarniczy* (May, 2006), p. 22.

43 Sovacool and Cooper (2008).

44 Charles Perrow, "The President's Commission and the Normal Accident," in David Sills, Charles Wolf, and Vivian Shelanski (eds.), *The Accident at Three Mile Island: The Human Dimension* (Boulder: Westview Press, 1981), pp. 73–84; and John Kemeny *et al.*, *The Need for Change: The Legacy of Three Mile Island* (Washington, D.C.: Report of the President's Commission on the Accident at Three Mile Island, Government Printing Office, 1979).

45 *Report of the President's Commission on the Accident at Three Mile Island* (1979), available at http://www.pddoc.com/tmi2/kemeny/causes_of_the_accident.htm/.

46 Sovacool and Cooper (2008).

47 Lovins and Lovins (1982).

48 US Government Accountability Office, *Nuclear Regulatory Commission: Oversight of Nuclear Power Plant Safety Has Improved, But Refinements Are Needed* (Washington, D.C.: US GAO, September 2006, GAO-06-1029).

49 L.C. Cadwallader, *Occupational Safety Review of High Technology Facilities* (Idaho Falls, ID: Idaho National Engineering and Environmental Laboratory, January 2005, INEEL/EXT-05-02616).

50 US Government Accountability Office, *Nuclear and Worker Safety: Actions Needed to Determine the Effectiveness of Safety Improvement Efforts at NNSA's Weapons Laboratories* (Washington, D.C.: US GAO, October 2007, GAO-08-73).

51 US Government Accountability Office (2008).

52 US Government Accountability Office (2006).

53 Bradford (2009).

54 Richard Webster and Julie LeMense, "Spotlight on Safety at Nuclear Power Plants: The View from Oyster Creek," *Pace Environmental Law Review* 26 (2009), p. 388.

55 Massachusetts Institute of Technology, *The Future of Nuclear Power: An Interdisciplinary MIT Study* (2003), p. 4, available at http://web.mit.edu/nuclearpower/pdf/nuclearpower-summary.pdf/.

56 Bernard Papin and Patrick Quellien, "The Operational Complexity Index: A New Method for the Global Assessment of the Human Factor Impact on the Safety of Advanced Reactor Concepts," *Nuclear Engineering and Design* 236 (2006), pp. 1113–1121.

57 Christian Parenti, "Nuclear Power Is Risky and Expensive," in Peggy Becker (ed.), *Alternative Energy* (New York: Gale, 2010), pp. 50–55.

58 Robert Alvarez, Jan Beyea, Klaus Janberg, Jungmin Kang, Ed Lyman, Allison Macfarlane, Gordon Thompson, and Frank N. von Hippel, "Reducing the Hazards from Stored Spent Power-Reactor Fuel in the United States," *Science and Global Security* 11 (2003), pp. 1–51.

59 Quoted in Sovacool and Cooper (2008).

60 David Lochbaum, *U.S. Nuclear Plants in the 21st Century: The Risk of a Lifetime* (Cambridge, MA: Union of Concerned Scientists, 2004), p. 5.

61 James R. Temples, "The Politics of Nuclear Power: A Subgovernment in Transition," *Political Science Quarterly* 95(2) (Summer, 1980), pp. 239–260.

62 John Byrne and Steven M. Hoffman, "The Ideology of Progress and the Globalisation of Nuclear Power," in John Byrne and Steven M. Hoffman (eds.), *Governing the Atom: The Politics of Risk* (London: Transaction Publishers, 1996), pp. 11–46.

63 Greenpeace (2008).

64 Frank von Hippel, "Managing Spent Fuel in the United States: The Illogic of Reprocessing," in Henry D. Sokolski (ed.), *Falling Behind: International Scrutiny of the Peaceful Atom* (Washington, D.C.: Nonproliferation Education Center, 2008), pp. 159–219; and B. Banerjee and N. Sarma, *Nuclear Power in India: A Critical History* (New Delhi: Rupa & Company, 2008).

65 Banerjee and Sarma (2008).

66 *Ibid.*

67 *Ibid.*

68 Greenpeace, *Nuclear Power: A Dangerous Waste of Time* (Amsterdam: Greenpeace International, 2009), p. 11.

69 Matthew L. Wald, "In Finland, Nuclear Renaissance Runs into Trouble," *New York Times*, May 29, 2009.

70 *Ibid.*

71 Greenberg (1996).

72 A.E. Farrell, H. Zerriffi, and H. Dowlatabadi, "Energy Infrastructure and Security," *Annual Review of Environment and Resources* 29 (2004), pp. 421–469.

73 Bradford (2009).

74 Yung-Tsan Jou, Tzu-Chung Yenn, Chiuhsiang Joe Lin, Chih-Wei Yang, and Chih-Cheng Chiang, "Evaluation of Operators' Mental Workload of Human–System Interface Automation in the Advanced Nuclear Power Plants," *Nuclear Engineering and Design* 239 (2009), pp. 2537–2542.

75 Smith (2009), p. 181.

[76] Greenberg (1996).

[77] Shahla M. Werner, "Nuclear Energy Too Risky When Efficiency Works," *Milwaukee Journal Sentinel*, May 9, 2009.

[78] Helen Caldicott, "The Dangers of Nuclear Power," *Australian Financial Review* (January 18, 2002), p. 18.

[79] Louis J. Sirico, "Stopping Nuclear Power Plants: A Memoir," *Villanova Environmental Law Journal* 21 (2010), pp. 35–44.

[80] Alvarez *et al.* (2003).

[81] Farrell *et al.* (2004), p. 454.

[82] The Alvarez *et al.* (2003) study produced quite a controversy. The US Nuclear Regulatory Commission responded in "Nuclear Regulatory Commission (NRC) Review of 'Reducing the Hazards from Stored Spent Power-Reactor Fuel in the United States,'" *Science and Global Security* 11 (2003), pp. 203–211, that the study (1) exaggerated the probability of a spent-fuel-pool fire; (2) overestimated the release of 30-year half-life cesium-137; (3) overestimated the damage from the release; and (4) underestimated the costs of moving to dry-storage casks a large fraction of the older spent fuel currently in spent-fuel pools. Alvarez *et al.* responded in Robert Alvarez, Jan Beyea, Klaus Janberg, Jungmin Kang, Ed Lyman, Allison Macfarlane, Gordon Thompson, and Frank von Hippel, "Response by the Authors to the NRC Review of Reducing the Hazards from Stored Spent Power-Reactor Fuel in the United States," *Science and Global Security* 11 (2003), pp. 213–223. They retorted that the NRC's critique in each of those four areas evaporated upon detailed inspection: (1) on probabilities, the NRC restated some of Alvarez *et al.*'s observations as if they had said the opposite; (2) on cesium-137 releases from a spent-fuel fire, the NRC adopted the lower end of Alvarez *et al.*'s uncertainty range by simply assuming that a fire would not spread from recently discharged to older spent fuel; (3) on damage, the NRC asserted that projections of the future population density around US reactors used in a 1997 study done for it were unrealistically high without offering an alternative; and (4) on costs, the NRC not only argued incorrectly that Alvarez *et al.* had neglected certain costs of removing 80% of the spent fuel currently in spent-fuel pools but also ignored lower-cost options that Alvarez *et al.* had urged it to examine as well.

[83] Webster and LeMense (2009), pp. 365–390.

[84] Nuclear Energy Agency, *Nuclear Education and Training: Causes for Concern* (Paris: OECD, 2000).

85   *Ibid.*, p. 14.

86   US Department of Energy, *A Roadmap to Deploy New Nuclear Power Plants in the United States by 2010* (2001), p. 7.

87   Joseph Romm, *The Self-Limiting Future of Nuclear Power* (Washington, D.C.: Center for American Progress Action Fund, June 2008).

88   Paul W. Benson and Fred Adair, "Nuclear Revolution," *Public Utilities Fortnightly* (July 1, 2008), p. 14.

89   Robin Pagnamenta, "Skills Crisis Looming in UK Nuclear Industry," *The Times (London)*, November 5, 2007.

90   Cogent Industries, *Skills for the Sciences: Nuclear* (London: Cogent, 2008), available at http://skillsreport.cogent-ssc.com/industries-nuclear.htm/.

91   International Atomic Energy Agency, *Analysis of Uranium Supply to 2050* (2001), pp. 34–39.

92   *Ibid.*, p. 5.

93   Paul Wenske, "Uranium Supply Questions: Finding Fuel for an Expanded Fleet," *EnergyBiz Insider* (September/October, 2008), p. 16.

94   Source: International Atomic Energy Agency, *World Distribution of Uranium Deposits (UDEPO) with Uranium Deposit Classification: 2009 Edition* (Vienna: IAEA, October 2009, IAEA-TECDOC-1629).

95   Wenske (2008).

96   Sovacool (2010).

97   *Ibid.*

98   *Ibid.*

99   *Ibid.*

100   International Atomic Energy Agency (2009).

101   Benjamin K. Sovacool, "Coal and Nuclear Technologies: Creating a False Dichotomy for American Energy Policy," *Policy Sciences* 40 (2007), pp. 116–117.

102   Gavin M. Mudd and Mark Diesendorf, "Sustainability of Uranium Mining and Milling: Toward Quantifying Resources and Eco-Efficiency," *Environmental Science & Technology* 42 (2008), pp. 2626–2629.

103   Edward D. Kee, "Nuclear Fuel: A New Market Dynamic," *Electricity Journal* 20(10) (December, 2007), pp. 54–64.

104   Oxford Research Group, "Energy Security and Uranium Reserves," *Secure Energy: Options for a Safer World Factsheet 4* (2006), available at http://www.oxfordresearchgroup.org.uk/publications/briefing_papers/energy_security_and_uranium_reserves_secure_energy_factsheet_4/.

[105]   Jeremy Whitlock, "Canadian Nuclear FAQ — Uranium" (2010), available at http://www.nuclearfaq.ca/cnf_sectionG.htm/ (accessed February 24, 2010).

[106]   A.D. McKay and Y. Miezitis, *Australia's Uranium Resources, Geology and Development of Deposits* (Canberra: AGSO Geoscience Australia, 2007, Mineral Resource Report 1).

[107]   Helen Caldicott, *Nuclear Power Is Not the Answer* (New York: The New Press, 2006), p. 318.

[108]   Antony Froggatt, "Nuclear Self-Sufficiency — Can Nuclear Power Pave the Road Towards Energy Independence?," Presentation to the "Towards a Nuclear Power Renaissance" Conference in Potsdam, Germany, March 4–5, 2010.

[109]   Luc Gagnon, "Civilization and Energy Payback," *Energy Policy* 36 (2008), pp. 3317–3322.

[110]   Luc Gagnon, *Electricity Generation Options: Energy Payback Ratio* (Montreal: Hydro Quebec, July 2005, 2005G185-A).

[111]   Scott W. White and Gerald L. Kulcinski, "Birth to Death Analysis of the Energy Payback Ratio and CO_2 Gas Emission Rates From Coal, Fission, Wind, and DT-Fusion Electrical Power Plants," *Fusion Engineering and Design* 48 (2000), pp. 473–481.

4

Unfavorable Economics: "Too Costly to Matter"

In the 1970s, the Washington Public Power Supply System (WPPSS) began a construction program for seven new nuclear power plants.[1] Planners at the WPPSS believed that electricity requirements would grow by 5.2% each year well into the 1990s and started building nuclear power plants to meet their projections. At the same time, however, a massive backlog of nuclear power plant orders after the 1973 oil crisis caused a severe shortage of skilled nuclear engineers and architects; 69 plants were ordered in 1973 and 1974. Problems of plant design, poor craftsmanship, and labor strikes caused even longer delays, forcing five-year construction estimates to extend into 10- or 12-year periods. One WPPSS project started in 1970 was not finished until 1984, and the WPPSS annual report in 1981 projected that US$23.7 billion was needed to complete one of its plants after US$5 billion had already been expended; all the while, electricity growth dropped significantly below original projections. By the mid-1980s, the WPPSS faced financial disaster and all but one of its plants was canceled, leading to the largest municipal bond default in the history of the US.

The entire experience came to be called the "WHOOPS" fiasco, as a play on the WPPSS acronym, and it is an enduring lesson of the risk associated with investing in large power plants. Consumers across the Northwest of the United States are still paying for WHOOPS in their monthly electricity bills.[2] The experience with WHOOPS, however, is not necessarily confined to the past. As this chapter will show, nuclear plants are grotesquely capital-intensive and expensive at almost all stages of the fuel cycle, especially construction, fuel reprocessing, waste storage, decommissioning, and research on new nuclear technology. These exceptionally high costs are connected, in part, to the history of nuclear power itself, as neither the US nor France — two countries largely responsible for developing nuclear power — pursued nuclear power generators for their cost-effectiveness. The chapter documents that modern nuclear power plants are therefore prone to at least seven different types of expenses: construction, reprocessing, storage, decommissioning, fuel, security, and research costs.

Taken together, the economic costs of new nuclear power plants will likely act as a major impediment to any nuclear renaissance. As the physicist Amory Lovins and his colleagues recently mused, nuclear power — once claimed to be "too cheap to meter" — may now be "too costly to matter."[3] Moreover, as will be noted in the following two chapters, despite enormous government subsidies, nuclear plants still suffer from a host of other insidious and inescapable challenges related to land use, water consumption, greenhouse gas emissions, and the proliferation of weapons of mass destruction.

Construction and Operating Costs

As the WHOOPS fiasco illustrates, nuclear power plants have long construction lead times and meet with a plethora of uncertainties during the construction process, making planning and financing difficult, especially when the balance of supply and demand for electricity can change rapidly within a short period of time. Long construction times become significant because costs mount quickly during construction delays. Halting construction of a nuclear power plant for two years adds about 15% to the final cost of electricity. The nuclear demonstration plant at Shippingport, Pennsylvania, for instance, was budgeted at US$48 million in the early 1950s, but ended up costing US$84 million by the time it was completed on December 2, 1957, and that excludes government subsidies and research expenses.[4]

What explains the proclivity of nuclear reactors to cost overruns and delays? That is, what makes new plants so expensive? Part of the explanation lies in their capital-intensive nature. Nuclear power plants are more like cathedrals than cellular telephones: they must be individually designed and built on a site-by-site basis. Modern nuclear plants are the most expensive and capital-intensive structures ever built, and they are the lynchpin of an industry that is already the most capital-intensive in most countries. Luis Echávarri, head of the Nuclear Energy Agency (NEA), reports that initial construction of a new nuclear reactor consumes close to 60% of the project's total investment, compared to about 40% for coal power plants and 15% for natural gas power plants.[5]

On top of their capital intensity, the approval process before construction even begins in places like the US takes about 42 months.[6] Once approved, Table 1 shows that the average construction time for all global nuclear power plants built from 1976 to 2007 was greater than seven years. In some particular places, construction lead times were even worse: the average lead time for nuclear power plant construction in the 1980s in Europe was more than 132 months.[7] Table 2 shows how the last four reactors built in France, two units in Chooz and two in Civaux, took 126 months to build.[8] In Germany, it took an average of 76–110 months to build a nuclear power plant from 1965 to 1989; in Japan, 44–61 months from 1965 to 2004; and in Russia, 57–89 months from 1965 to 1993.[9]

Table 1: Average Global Construction Time for Nuclear Reactors, 1976–2007[10]

Period	Number of Reactors	Average Construction Time (months)
1976–1980	86	74
1981–1985	131	99
1986–1990	85	95
1991–1995	29	104
1996–2000	23	146
2001–2005	20	64
2006	2	77
2007	3	80

Table 2:   Construction Times for French Pressurized Water Reactors, 1971–2007[11]

Sites	Reactor Type	Reactor Size (MWe)	Number Built	Constructed Between	Mean Construction Time (months)
Bugey and Fessenheim	PWR Westinghouse license	900	6	1971–1979	63
Blaye, Dampierre, Gravelines, and Tricastin	As CPO	900	18	1974–1985	65
Chinon, Cruas, and St. Laurent	As CP1	900	10	1976–1987	67
Flamanville, Paluel, and St. Alban	PWR Westinghouse license	1,300	8	1977–1986	78
Belleville, Cattenom, Golfech, Nogent, and Penley	P4 Westinghouse	1,300	12	1979–1993	90
Chooz and Civaux	PWR French design	1,500	4	1984–1999	126
Flamanville	EPR	1,600	2	2007–?	—

These long construction times place nuclear power plants at great risk to unforeseen changes in electricity demand, interest rates, availability of materials, severe weather, and labor strikes, all of which contribute to severe cost overruns. One study estimated nuclear power plant construction costs from 1966 to 1977, when most light water reactors in the US were built, and found that in every case plants cost *twice as much* as expected or more.[12] The results of that study are presented in Table 3. The quoted cost for these 75 plants (when updated to 2009 USD) was US$89.1 billion, but the real cost was a monumental US$283.3 billion.[13] These numbers paint only a partial picture, as they refer to only those plants that were actually built; another 130 reactors had US$20 billion funneled into

Table 3: Estimated and Actual Costs of 75 Nuclear Power Plants in the United States

Plant	Estimated Costs at Start of Construction (US$ million as of 1990)	Realized Cost (US$ million as of 1990)	Plant	Estimated Costs at Start of Construction (US$ million as of 1990)	Realized Cost (US$ million as of 1990)
Arkansas Nuclear 1	$375	$624	McGuire 1	$414	$1,299
Arkansas Nuclear 2	$460	$1,081	McGuire 2	$472	$1,269
Beaver Valley 1	$513	$1,176	Millstone 2	$474	$936
Beaver Valley 2	$913	$4,099	Millstone 3	$1,046	$3,998
Braidwood	$762	$2,723	Nine Mile Point 2	$1,008	$5,281
Browns Ferry 1	$303	$876	North Anna 1	$515	$1,555
Browns Ferry 2	$227	$657	North Anna 2	$445	$932
Browns Ferry 3	$227	$657	Palisades	$294	$422
Brunswick 1	$430	$718	Palo Verde 1	$1,234	$4,185
Brunswick 2	$352	$933	Palo Verde 2	$920	$2,291
Byron 1	$741	$2,518	Peach Bottom 2	$532	$1,418
Byron 2	$552	$2,072	Peach Bottom 3	$423	$560
Callaway	$1,136	$2,999	Perry 1	$981	$3,729
Calvert Cliffs 1	$357	$1,142	Rancho Seco	$389	$876
Calvert Cliffs 2	$287	$765	River Bend 1	$718	$4,091
Catawba 1	$559	$2,074	Salem 1	$462	$1,829
Clinton	$710	$4,058	Salem 2	$378	$1,497
Cooper	$378	$1,053	San Onofre	$1,134	$3,343
Crystal River 3	$362	$948	San Onofre 3	$1,056	$2,078

(Continued)

Table 3: *(Continued)*

Plant	Estimated Costs at Start of Construction (US$ million as of 1990)	Realized Cost (US$ million as of 1990)	Plant	Estimated Costs at Start of Construction (US$ million as of 1990)	Realized Cost (US$ million as of 1990)
Davis-Besse 1	$484	$1,359	Sequoyah 1	$524	$1,560
Diablo Canyon 1	$445	$3,750	Sequoyah 2	$429	$1,276
Diablo Canyon 2	$459	$2,333	Shoreham	$300	$4,139
Donald C. Cook 1	$657	$1,303	St. Lucie 1	$365	$1,130
Duane Arnold	$340	$716	St. Lucie 2	$893	$1,876
Edwin I. Hatch 1	$417	$951	Surry 1	$419	$761
Edwin I. Hatch 2	$653	$922	Surry 2	$329	$437
Fermi 2	$596	$3,783	Susquehanna 1	$1,320	$2,654
Fort Calhoun 1	$222	$520	Susquehanna 2	$753	$2,274
Grand Gulf 1	$1,105	$3,473	Three Mile Island 1	$323	$1,008
Harris 1	$898	$3,999	Three Mile Island 2	$668	$1,287
Hope Creek	$1,592	$4,598	Trojan	$582	$1,145
Indian Point	$477	$859	Virgil Summer 1	$630	$1,707
Joseph M. Farley 1	$387	$1,463	Waterford 3	$617	$3,303
Joseph M. Farley 2	$406	$1,228	Wolf Creek 1	$1,143	$2,835
Kewaunee	$297	$559	WPSS 2	$786	$4,008
LaSalle 1	$715	$1,918	Zion 1	$593	$768
LaSalle 2	$532	$1,255	Zion 2	$430	$752
Limerick 1	$921	$3,980	Total	$45,247	$144,650

them before they were abandoned between 1972 and 1984.[14] The Shoreham Nuclear Power Plant adjacent to the Wading River in East Shoreham, New York, cost ratepayers US$6 billion, but was closed by protests in 1989 before the plant could generate a single kilowatt-hour of electricity.[15] Across the border in Canada, delays and cost overruns on nuclear power plants accounted for C$15 billion of the nearly C$20 billion of "stranded debt" created by Ontario Hydro.[16] Almost halfway around the world, the canceled Bataan nuclear power plant near Manila in the Philippines ended up costing ratepayers US$2.3 billion, even though it was never switched on after social protests convinced the government to moth-ball it. In India, Table 4 reveals that the completed construction costs for its past ten reactors have been substantially overbudget.

Many other factors contribute to the prevalence of cost overruns in nuclear power plant construction. Atomic energy was not technically or economically feasible when it was embraced by the US in 1946, meaning it did not originate as an invention from the marketplace, nor did it evolve with any particular consumers in mind. The US government had to commit to 11 years of intensive research before nuclear power was even successfully demonstrated.[18] The Dwight D. Eisenhower administration decided to develop nuclear power plants in the 1950s for entirely political reasons, seeking to demonstrate a positive aspect of nuclear energy after

Table 4: Original and Revised Cost Estimates for Indian Nuclear Reactors, 1972–2006[17]

Station	Original Cost Estimate (10 million rupees)	Revised Cost (10 million rupees)	Year Completed	Relative Cost Increase (%)
RAPS I	339.5	732.7	1972	216
RAPS II	5,816	1,205.4	1980	176
MAPS I	617.8	1,188.3	1983	192
MAPS II	706.3	1,270.4	1985	179
NAPS I and II	2,098.9	7,450	1989 and 1991	354
Kakrapar I and II	3,825	13,350	1992 and 1995	349
Kaiga I and II	7,307.2	28,960	1999 and 2000	396
RAPS III and IV	7,115.7	25,110	2000	353
Tarapur III and IV	24,275.1	62,000	2006	255

World War II and instigate a technology race with the Soviet Union. Even after demonstration plants were completed, the federal government had to commit another US$20 billion in research and severely limit liability for electric utilities before they would consider operating nuclear plants.[19] In France, Charles de Gaulle promoted nuclear power plants as a mechanism to reconstruct French national identity. The end of World War II had left France humiliated and defeated, and the country lacked infrastructure, food, and political influence. French technical and scientific experts offered a solution to this dilemma by linking technological advancement to French "radiance," or identity. Nuclear technology was seen by French policymakers as a way to simultaneously rebuild French infrastructure and re-establish its role as a world leader. After the creation and demonstration of the atomic bomb, "nuclear technology became a quintessential symbol of modernity and national power."[20] French policymakers desired to promote nuclear power so much that one fourth of household income throughout the country went to the construction of the first 50 nuclear plants.[21] In both the American and French cases, the government created a market for nuclear power, rather than the other way around.

Also, many cost estimates are made by project sponsors and vendors with a penchant for underestimating the length of delays, setting contingencies too low, and ignoring risks relating to currency exchange, inflation, and public approval — factors exacerbated by a strong interest in presenting nuclear projects in terms as favorable as possible.[22] One survey of the real construction costs of nuclear power facilities at 16 operational reactors in Canada, China, Japan, the UK, and the US found that many of the construction costs quoted by industry representatives, promotional bodies, plant vendors, and utilities were unreliable, inconsistent, and conservative. These estimates excluded interest during construction and borrowing fees as well as the expense of decommissioning and fuel storage.[23] Another July 2008 survey from two energy consultants found that the construction costs quoted by industry suppliers were misleadingly incomplete because they excluded expenses related to procuring land and building cooling towers and switchyards, interest during construction, inflation, cost overruns, and contingency fees.[24] The study noted that, when these excluded items are included, they can often double the price of a nuclear power plant.

Of course, these examples are all historical. What do cost estimates for future nuclear plants look like today when not made by self-interested vendors and operators? Not good. Researchers from the Keystone Center, a nonpartisan think tank, consulted with representatives from 27 nuclear power companies and contractors, and concluded in June 2007 that the cost for building new reactors would be between US$3,600 and US$4,000 per installed kilowatt (with interest).[25] The projected operating costs for these plants would be remarkably expensive: 30 cents/kWh for the first 13 years until construction costs are paid, followed by 18 cents/kWh over the remaining lifetime of the plant. Just a few months later, in October 2007, Moody's Investors Service projected even higher operating costs — an assessment easily explained by the quickly escalating price of metals, forgings, other materials, and labor needed to construct reactors — and thus indicated that actual costs have "blown past" earlier estimates. They estimated total costs for new plants, including interest, at between US$5,000 and US$6,000 per installed kilowatt.[26] Florida Power & Light estimated the cost for building two new nuclear units at Turkey Point in South Florida to be US$8,000 per installed kilowatt, or a shocking US$24 billion in total. Progress Energy, in North Carolina, projected that a new plant would cost US$7,700 per installed kilowatt. Puget Sound Energy quoted a capital cost as high as US$10,000 per installed kilowatt. Figure 1 puts nuclear construction costs in context compared to other sources of electricity supply, such as renewables, and shows that nuclear power plants are the second most expensive facilities to build on the market (after solar photovoltaics).

Inflated construction costs are not limited to the US. In Canada, Ontario Power reported a staggering US$10,800 per installed kilowatt, or US$26 billion in total, for a new plant.[28] Such numbers mean that a single project would wipe out the province's nuclear-power expansion budget for the next 20 years, leaving no money for at least two more multibillion-dollar refurbishment projects. As one energy consultant in Toronto exclaimed, "it's shockingly high." Construction costs as high as US$11,000 per installed kilowatt for Finland, Canada, United Arab Emirates, and Turkey have also been quoted.[29] Put in perspective, from October 2000 to October 2007, material, labor, and engineering costs for nuclear power plants jumped 185%, meaning a plant that cost US$4 billion to build in

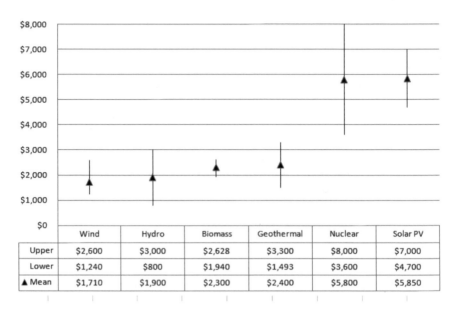

Figure 1: **Upper Range, Lower Range, and Mean Capital Cost for Nuclear and Renewable Power Plants (2007 US$/installed kW)**[27]

2000 would have cost US$11.4 billion in 2007. The Power Capital Costs Index, which tracks the expense of power plant materials and components, showed an increase in nuclear costs by a factor of 2.3 between 2000 and 2008; moreover, these costs were only for engineering and procurement, and still excluded other "soft costs" such as contingency fees, risk premiums, land use fees, and permit fees.[30] Another survey of the overnight construction costs for nine light water reactors recently built in South Korea and Japan documented that the cost of building new plants would likely be significantly higher (see Figure 2).[31] The study warned that constraints in the manufacturing of nuclear components, lack of skilled construction teams, and long lead times meant that a new nuclear plant would cost twice as much per installed kilowatt as previously quoted. Even with a carbon tax of US$30 per ton on carbon dioxide and requirements for carbon sequestration, the study concluded that new nuclear power plants would have no economic advantage over fossil-fueled or renewable energy technologies.

Figure 2:   Construction Workers Inspect the Foundation of One of the Reactors Being Built at the Kori Nuclear Power Plant in South Korea

Note: The facility is so large that the workers can only be seen as small specks in the bottom left of the photograph.

Furthermore, researchers at Georgetown University, the University of California at Berkeley, and the Lawrence Berkeley National Laboratory assessed financial risks for advanced nuclear power plants utilizing a three-decade historical database of delivered costs from each of 99 conventional nuclear reactors operating in the US.[32] Their assessment found a significant group of plants with extremely high costs: 16% in the more-than-8 cents/kWh category. The authors then applied their findings to the construction of new, advanced reactors. They pointed out two unique attributes of advanced nuclear power plants that make them prone to unexpected increases in cost: (1) their dependence on operational learning, a feature not well suited to rapidly changing technology and market environments subject to local variability in supplies, labor, technology, public opinion, and the risks of capital cost escalation; and (2) the difficulty in standardizing new nuclear units, or the idiosyncratic problems of relying on large generators whose specific site requirements do not allow for mass production. Past technology development patterns suggest that many high-cost surprises will occur in the planning and deployment process for new nuclear units. These "hidden" but inevitable cost overruns may be one reason why most investors have shied away from financing Generation IV reactors.

These capital and operating costs are so high that they deter most investors and utilities (and explain, as Chapter 8 discloses, why nuclear power needs profligate government subsidies in order for investors to embrace it). Costs upward of US$10 billion to build a new reactor are equivalent to 75–100% of the capitalization of assets for an entire electric utility. As Lew Hay, Chairman and CEO of Florida Power & Light, recently noted:

> If our cost estimates are even close to being right, the cost of a two-unit [nuclear] plant will be on the order of magnitude of $13 to $14 billion. That's bigger than the total market capitalization of many companies in the U.S. utility industry and 50 percent or more of the market capitalization of all companies in our industry with the exception of Exelon.[33]

To build the 20 potential new nuclear power plants currently being considered in the US, US$300 billion in investments would be needed.

This amount is double the estimated cost to repair all of the bridges in the US road transportation system, is more than the cost of bringing all American public school buildings into conformity with building standards, and is also larger than the combined gross domestic product of 35 individual states (including Maryland, Colorado, Missouri, Alabama, and South Carolina).[34] One 2009 assessment of nuclear energy in the US noted that the cost of building 100 new reactors, rather than pursuing cheaper methods such as energy efficiency, would amount to an extra US$1.9–$4.4 trillion over the life of the reactors.[35]

These outrageously high costs for nuclear power explain why most plants are unable to receive financing from the private market. The *Wall Street Journal* has indicated that electric utilities and suppliers have been "hobbled" by the financial crisis and are unable to finance new projects, especially large nuclear reactors. Interest rates are unfavorable, and some utilities in deregulated markets currently face the risk of credit downgrades.[36] The financial risks of new nuclear reactors have led private firms and investors to indicate that it will be difficult, if not impossible, to sell bonds to support projects in capital markets.[37]

Reprocessing Costs

In the early days of the nuclear era, plutonium was considered a possible "silver bullet" solution to the world's energy problems. It was thought that the continuous burning, breeding, and recycling through a collection of reactors would eliminate the need for uranium mining and enrichment and one day replace fossil fuels altogether. This idea pushed two related research programs in the early 1950s: the separation of plutonium from spent uranium oxide fuel, known as "reprocessing"; and the development of fast reactor systems, utilizing a process known as "breeding." When nuclear engineers initially designed American reactors, they took these views into consideration and expected that the plutonium from spent fuel would be recycled at reprocessing centers or removed and reused in fast neutron reactors.

The option to reuse plutonium at fast neutron reactors or fast breeder reactors was rejected by political overseers on national security grounds. Because of the link between plutonium and nuclear weapons, the potential application of fast breeders led to concerns that nuclear power expansion would usher in an era of uncontrolled weapons proliferation. The US signed the Nuclear Non-Proliferation Treaty in 1968 partially to address this issue, but India's unexpected test of a nuclear device in 1977 took the US by surprise and culminated in President Jimmy Carter's nonproliferation policy which banned civilian reprocessing of nuclear fuel.[38]

The US federal government did begin efforts on commercial reprocessing of nuclear waste in 1966 at a facility in West Valley, New York, but the operation ended in disaster.[39] The plant was repeatedly criticized for lax security measures and for exposing employees to dangerously high doses of radiation, exceeding federal regulations dictated by the Occupational Safety and Health Administration, which was established in 1970.[40] In addition, the project ran into insurmountable logistical problems. The cost of reprocessing was originally estimated to be US$15 million, but was later reported to be US$600 million; the probability of a major earthquake in the area was deemed too great a risk to justify continued operation; and in practice the reprocessing plant was far less efficient than engineers had originally estimated.[41] After reprocessing only 640 tons of spent fuel, while accumulating more than 600,000 gallons of high-level waste, the facility was closed in 1972. It was not until 2002 that the West Valley facility was stabilized to the point that it could be safely decommissioned. However, remaining cleanup was estimated in 2008 to cost an additional US$5 billion and take another 40 years.[42]

Newer modes of reprocessing, such as mixed oxide (MOX) fuel and uranium extraction plus (UREX+) reprocessing, are also expensive (and face daunting problems).[43] MOX fuel reprocessing produces dangerous levels of plutonium waste that can be used for weapons, meaning facilities must be guarded and nuclear fuel stored; these challenges are discussed in greater detail in Chapter 6. The quality of recycled fuel significantly decreases the more it is reprocessed. A reduction in quality occurs each time fuel is reprocessed and recycled; and as fuel quality degrades, more energy is needed to enrich fuel rods, which makes the fuel even more dangerous due to greater emissions of neutron and gamma radiation, which in turn lead to higher overall burn-up rates and drastically less efficient fuel. Reactors cannot run on entirely recycled fuel. The industry standard is 30% MOX fuel and 70% fresh uranium. Plants still need significant supplies of natural uranium that must be mined from depleting stores of diminishing levels of quality ore, a problem detailed in Chapter 3. Reprocessing capacity is also currently constrained. The two largest reprocessing facilities can process only 320 out of 2,500 tons of waste per year combined — a mere 12.8% of the nuclear waste created in Europe annually. Since plants must shut down to load MOX fuel, reprocessing has led to loading problems as operators are reluctant to power down units that will have to be offline for 37 days to refuel.

The UREX+ method is both unproven and absurdly expensive. The US Department of Energy (DOE) estimated in 1999 that it would cost US$279 billion over a 118-year period to fully implement a reprocessing and recycling program for the existing inventory of US spent fuel relying on UREX+.[44] The National Academies concurred, and noted in 2008 that "there is no economic justification for going forward with [a UREX+] program at anything approaching a commercial scale. . . . [UREX+] is [not] at a stage of reliability and understanding that would justify commercial-scale construction at this time. Significant technical problems remain to be solved."[45] The extra costs involved with the separation and transmutation of waste were found to be US$100 billion *more* than simply storing it in a repository.

The US government contracted with Areva to build a test MOX fuel fabrication plant to reprocess 34 tons of plutonium, but its cost has ballooned from US$1 billion to US$3.5 billion and it is more than five years behind schedule. The nonpartisan Congressional Budget Office warned that the Global Nuclear Energy Partnership (GNEP)'s plan to reprocess spent fuel would cost 25% more than a wide range of other storage and direct disposal options.[46] They projected that reprocessing spent nuclear fuel would cost US$585–$1,300 per kilogram, an upper amount more than twice as much as direct disposal. For the roughly 2,200 metric tons of spent fuel produced each year in the US, the study projected that employing reprocessing as an alternative would likely cost at least an extra US$5 billion.[47] Researchers at the CEA (*Commissariat à l'énergie atomique*) in France looked at five Generation IV reactors and theoretical models of their associated fuel cycles from 2000 to 2150. They found that Generation IV reactors entailed much higher reprocessing and disposal costs compared to conventional recycling and fuel disposal, and estimated that the

Generation IV pathway would cost 30–45% more than business as usual.[48] Lastly, both types of reprocessing create their own waste, including elements such as plutonium-241 and americium-241 that would still need to be separated and stored in a centralized repository, adding to the expense.

Worldwide, about half of the plutonium being separated is simply being stockpiled at reprocessing plants. As of 2005, this global stockpile of separated plutonium exceeded 250 tons, or enough to make more than 30,000 nuclear weapons. Interestingly, while the stockpile is enough to make an arsenal of weapons, it could only fuel the world's fleet of nuclear reactors in 2008 for less than one year.

Waste Storage Costs

The cost of temporarily and permanently storing nuclear waste is also prohibitively expensive. As of 2007, not a single country had yet completed the construction of a long-term geologic repository for nuclear waste. The responsibility for permanently storing America's nuclear waste falls exclusively to the federal government, but it is clearly failing in its role. The Nuclear Waste Policy Act (NWPA) of 1982 obligated utilities to pay a fixed annual fee — one tenth of a cent for every kilowatt-hour from nuclear generation — that would be collected in a Nuclear Waste Fund to cover the costs of waste disposal. In return, the federal government and the DOE were required to take and dispose of spent nuclear fuel in a permanent geologic repository beginning in 1998. Pursuant to the NWPA, nine states were initially identified as potential sites for long-term repositories, but, for political reasons, regulators quickly abandoned all but one of these sites: Yucca Mountain in Nevada.

Ironically, scientists had deemed Yucca the least optimal of the nine sites. The National Academies of Sciences reported that it had the greatest risk of releasing dangerous levels of radiation. Nonetheless, because it was the only alternative, the federal government began funding a permanent storage facility at Yucca Mountain in 1985.[49] By 2008, the project had already cost US$13.5 billion and was some 20 years behind schedule, underfunded, and, according to Nevada Senator Harry Reid, who opposed it, "a dying beast." Even if it had been completed on schedule, Yucca would have only enough space for 70,000 tons of spent fuel, leaving 35,000 tons of radioactive waste that would require storage by 2035, assuming the existing fleet of nuclear reactors continued to operate. The Congressional Budget Office noted in the same year that they expected the construction of Yucca Mountain to take another century and exceed US$57 billion. Just one year later, the DOE offered an updated estimate that the cost of building and operating Yucca Mountain would exceed US$96 billion; and this staggering price tag only covered the expense of building the facility and transporting nuclear waste until 2133. In 2009, the US Government Accountability Office (GAO) projected that building a separate repository to dispose of 153,000 metric tons of waste would cost US$41–$67 billion over a 143-year period.[50] Perhaps because of these reasons, the Barack Obama administration announced in early 2010 that they intended to revoke the license application for Yucca Mountain and that a new appropriate site will be identified by a recently appointed Blue Ribbon Commission on America's Nuclear Future.

Worried that the government would not meet its responsibility to build a permanent storage facility, several electric utilities operating commercial nuclear reactors went before the D.C. Circuit Court in 1996 to seek a ruling on the extent of the government's obligations under the NWPA. In that case, *Indiana Michigan Power v. DOE*, the court ruled that the government had to unconditionally accept waste by January 31, 1998. Without seeking a rehearing, the government nonetheless informed utilities that it would not accept the deadline. Facing growing quantities of nuclear waste and limited storage space, utilities responded and petitioned the US Court of Appeals for a writ of mandamus to require the federal government to begin accepting highly radioactive spent fuel from the utilities by the following January; again, the government refused. By 2006, about 20 utilities had suits pending against the DOE in the Court of Federal Claims for damages which could total in the tens of billions of dollars. By February 2008, the number of lawsuits pending against the DOE related to nuclear storage had jumped to 60, with the potential total liability rising every month.[51] Concomitantly, nuclear power operators and utility companies have successfully filed a suit against the DOE, recovering more than US$7 billion thus far in damages caused by the DOE's inaction. Ratepayers continue to pay one tenth of a cent per kilowatt-hour of nuclear electricity, as well as about US$500 million per year in judgments from the Department of Justice, to say nothing of the US$24 billion sitting in the Nuclear Waste Fund and the 71 other lawsuits (the total as of early 2010) against the DOE still to be decided.

The DOE has relied upon onsite storage as a stopgap remedy until the US finds a long-term solution to nuclear waste. As a result, about 30,000 tons of spent nuclear fuel are scattered in dry casks and storage pools in 34 states. Twenty-six reactors were projected to be out of pool storage space in 1998, and 80 were expected to reach maximum pool capacity by 2010. One ton of highly radioactive waste is generated for every four pounds of usable uranium, and each reactor consumes, on average, 32,000 fuel rods over the course of its lifetime. The costs of expanding onsite storage are, therefore, enormous, with each dry cask running at about US$35,000–$65,000 per ton[52] and other storage methods running at about US$80,000 per ton (or US$470,000–$750,000 per site per year).[53] Canada has seen the projected time for the construction of its centralized storage facility grow even longer. The federally sponsored Nuclear Waste Management Organization reported in 2006 that it will need more than 300 years to implement its approach to "containing" spent nuclear fuel at an expense of at least US$24 billion.

Regardless of whether the nuclear waste problem is resolved in favor of onsite or centralized storage, the costs will not be borne solely by this generation, or even by generations over the next millennium. Typically, a single nuclear plant will produce 30 tons of high-level waste each year, and this waste can be radioactive for as long as 250,000 years. Assuming just one tenth of that time (25,000 years), and assuming the cost of storing the 30 tons of nuclear waste created per year was just US$35,000 per ton, the lowest end of existing estimates, each nuclear plant in the US would assume an additional cost of US$26.3 billion on top of its already enormous price tag. Given that more than 430 nuclear power plants exist, the global cost of storing only their current waste would surpass US$11.3 trillion.

Decommissioning Costs

The price of energy inputs and the environmental costs of every nuclear power plant continue long after the facility has finished generating its last useful kilowatt of electricity. Both nuclear reactors and uranium enrichment facilities must be tediously decommissioned — a process that is freakishly expensive, time-consuming, dangerous for workers, and hazardous to the natural environment. As Chapter 2 noted, after a cooling-off period that may last as long as a century, reactors must be dismantled and cut into small pieces to be packed in containers for final disposal. While it will vary along with the technique and reactor type, the total energy required for decommissioning can be as much as 50% more than the energy needed for original construction.[54]

Indeed, every nuclear facility in operation now and every nuclear plant that will ever come online will eventually reach the end of its useful life, and will begin the long and arduous task of decommissioning — that is, returning the facility, its parts, and surrounding land to a safe enough level to be entrusted for other uses. This decommissioning process includes all of the administrative and technical actions associated with ceasing operations, removing spent or unused fuel, reprocessing or storing radioactive waste, deconstructing and decontaminating structures and equipment, shipping contaminated equipment offsite, and remediating the land, air, and water around the reactor site. In most cases, the decommissioning process costs anywhere from US$300 million to US$5.6 billion.[55] Because decommissioning involves the dismantling and transport of substantial amounts of radioactively contaminated material, it presents new opportunities for accidents or sabotage even beyond the useful generating cycle of the facility. Moreover, because decommissioning involves a substantial shift in the normal operating procedures of the facility, it risks the introduction of unforeseen human error at every step in the process.

In the US, there are currently 13 nuclear power plant units that have permanently shut down and are in some phase of the decommissioning process, but not a single one of them has completed it.[56] For example, Peach Bottom Unit 1 was shut down in October 1974, but will not even begin decommissioning until 2034. The Humboldt Bay nuclear facility was shut down in July 1976, but will not be completely decommissioned until 2012 or 2013. Zion Units 1 and 2 were permanently shut down in 1998, but the plant will not begin decommissioning until 2013. Furthermore, unless license extensions are granted, all licenses for commercial nuclear reactors in the US will expire by 2038 and more than

100 reactors will enter the decommissioning phase, requiring billions of dollars with little or no generating capacity to offset these costs.

The expense of decommissioning does add significantly to the final cost of nuclear electricity when not subsidized and included in cost estimates. After surveying the decommissioning costs for 26 countries (and a variety of reactor types and sizes), including dismantling of the reactor core, disposal and treatment of waste, site cleanup and land remediation, and project management, the World Energy Council calculated that decommissioning adds 2.5–4.2 extra cents per kilowatt-hour to the final price of nuclear electricity.[57]

Decommissioning at nuclear sites that have experienced a serious accident is far more expensive and time-consuming. Three Mile Island Unit 2, which shut down permanently after an accident in 1979, will not start the decommissioning process until 2014. Fuel rods at Chernobyl, the site of the world's deadliest nuclear accident to date, are still being removed and operators expect it to take until at least 2038–2138 before the power plant is completely decommissioned (see Figure 3).[58]

The decommissioning of uranium enrichment facilities — large complexes of buildings with thousands of pieces of equipment, enrichment

Figure 3:   Decommissioning at the Chernobyl Nuclear Power Plant near Kiev, Ukraine

cascades, piping, and electrical wiring — requires a precarious six-stage and very labor-intensive process: careful characterization of every square centimeter of each building, disassembly, removal of uranium deposits from process equipment, decontamination, melt refining and recycling of metals, and treatment of waste. The process generates its own low-level radioactive and hazardous waste, and can further contaminate soil and groundwater. The Capenhurst gaseous diffusion plant in the UK, decommissioned by British Nuclear Fuels Limited in 1994, required the entire facility to be treated with gaseous chlorine trifluoride (ClF_3) to remove deposits of uranium on equipment before every piece of the plant was extracted, cut up into pieces, and decontaminated using a series of aqueous chemical baths.

The decommissioning of three enrichment facilities in the US — all of the gaseous diffusion type, with one retired facility located near Oak Ridge, Tennessee; one operating facility near Paducah, Kentucky; and another retired one near Portsmouth, Ohio — will require the same ClF_3 treatment, because deposits of highly enriched uranium have become littered throughout the process buildings. This radioactive debris is accompanied by significant amounts of asbestos and polychlorinated biphenyls. This is probably why the National Research Council estimated that decommissioning the facilities will cost US$27.3–$67.2 billion, with an additional US$2–$5.8 billion to cover the disposal of a large inventory of depleted uranium hexafluoride, which must be converted to uranium oxide (U_3O_8).[59]

The US GAO recently surveyed how well the decommissioning process was going at these enrichment facilities, and found that the earliest it will be completed for all three plants is 2044. By then, the GAO warned that the cost of decommissioning, funded by taxpayers, will have exceeded the plants' revenues by at least US$4–$6.6 billion in 2007 dollars.[60] As of 2004, these plants — which are heavily contaminated with radioactive particles and large caches of spent hexafluoride fuel — still required extensive cleanup of 30 million square feet of space, miles of interconnecting pipes, and thousands of acres of land. The Nuclear Decommissioning Authority in the UK has also reported similar problems with the decommissioning of its units, the costs of which are now estimated to be more than £73 billion. One of the companies responsible for decommissioning in the UK, the state-owned British Nuclear Fuels Limited, reported £356 million of shareholder funds in 2001 but £35 billion in liabilities from decommissioning reprocessing plants, underscoring the immensity of cleanup costs.[61] France is expected to spend at least €65 billion on decommissioning its uranium enrichment facilities and some, such as Brennilis, will cost 20 times the original sum envisioned by developers.[62] Likewise, the cleanup and decommissioning of the Rum Jungle uranium mine in Australia cost the Australian government far more than it ever earned from the mine.[63]

Fuel Costs

Reliance on uranium fuel poses two types of economic costs: dependency and price volatility. Accidents, severe weather, and bottlenecks can all prevent uranium from being adequately distributed to nuclear facilities in desperate need of fuel.

In terms of dependency, investments in new nuclear plants would only make nuclear power countries dependent on foreign deposits of uranium in Africa, Russia, Canada, and Australia. Admittedly, the chances that Canada and Australia will band together to become the new "OPEC of uranium" are as unlikely as it sounds; but Kazakhstan, Namibia, Niger, and Uzbekistan together were responsible for more than 30% of the world's uranium production in 2006. Over the past several years, these countries have suffered from autocratic rule and political instability. It is not inconceivable to imagine a scenario in which unstable or hostile regimes controlling only 30% of the world's supply of uranium could nonetheless induce price spikes and volatility in uranium supplies that could have devastating consequences to the West.[64]

The entire nuclear fuel cycle is also dependent on incredibly long lead times and geographically separated facilities. The time needed to bring major uranium mining and milling projects into operation averages five or more years for exploration and discovery, with an additional 8–10 years for production. Moreover, uranium conversion facilities currently only operate in Canada, France, the UK, the US, and Russia. Consequently, it typically takes between five and seven years before uranium from the ground actually reaches a nuclear reactor.

Expected shortfalls in available fuel not only threaten future plants, but also contribute to the volatility of uranium prices at existing plants. Uranium fuel costs have shown recent escalation and volatility, which negatively impact the economics of nuclear projects, making them riskier than many existing alternatives. For example, Figure 4 depicts uranium spot prices from 1965 to 2009 and reveals at least three astronomical spikes: one in the 1960s and 1970s, resulting from a rush of countries trying to develop weapons and exhausting easily accessible reserves; a second one in the 1970s and 1980s, as a large number of commercial nuclear power plants began operation; and a third one post-2004, caused by an expected renaissance and constraints in supply. The cost of uranium, for instance, jumped from US$7.25 per pound in 2001 to US$47.25 per pound in 2006 — an increase of more than 600%. The NEA reported that 200 metric tons of uranium are required annually for every 1,000-MW reactor and that uranium fuel accounts for 15% of the lifetime costs of a nuclear plant; thus, price spikes and volatility can cost millions of dollars, affecting the profitability of a plant.[65] Uranium price volatility has been heavily influenced by the unexpected introduction of secondary supplies

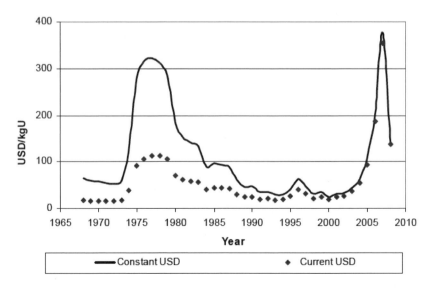

Figure 4:   Uranium Market Prices, 1965–2009[66]

and gluts in the market, connected in part to sudden increases in supply from canceled and shut-down reactors and the dilution of highly enriched uranium from surplus nuclear weapons.

Security Costs

Nuclear power is a form of energy supply that is unique from all others, due to its reliance on fission and its production of nuclear waste and weapons-grade materials. It therefore involves a long list of "hidden" security costs, one of the most pernicious relating to military expenditures. Although it is difficult to disaggregate the money spent defending military nuclear facilities from civilian ones, in the US the 2008 nuclear security budget was more than US$50 billion, and it exceeded all anticipated spending on international diplomacy and foreign assistance (US$39.5 billion) as well as on natural resources and the environment (US$33 billion). It was almost twice as large as the budget for general science, space, and technology (US$27.4 billion), and it was almost 14 times what the DOE allocated for all energy-related research and development.[67] Less than 2% (US$700 million) of the entire budget was spent on preparing for or responding to a nuclear accident or radiological attack.

Another assessment projected that the US DOE also spends approximately US$115 million each year on nuclear materials protection, control, and accounting programs. Of this amount, approximately US$58 million is spent on physical security and US$55 million on materials control and accounting. In addition, each year the DOE spends approximately US$320 million on guard forces, US$69 million on personnel security, US$46 million on research and development related to physical security, and US$21 million on the construction of security barriers and infrastructure.[68]

While data related to security costs are available for the US, they are, however, nonexistent for most Asian countries and also look only narrowly at physical security. The data do not account for other, indirect security costs such as the welfare loss caused by economic sanctions on nations violating the Nuclear Non-Proliferation Treaty (such as Iran and North Korea), or the military costs of protecting the sea lanes where uranium fuel and MOX plutonium traverse.

Research Costs

As a final category of costs, advanced nuclear research is expensive and highly uncertain. The National Research Council of the National Academies issued a highly critical assessment of the GNEP and the Generation IV program, arguing that its rapid deployment schedule entailed considerable financial and technical risks and prematurely narrowed the selection of acceptable reactor designs.[69] The report also faulted the DOE for not seeking sufficient independent peer reviewers for projects and for failing to adequately address waste management challenges. For example, because higher-temperature reactors tend to burn up more of their fuel at faster rates, they operate less efficiently than conventional units and result in more radioactive waste per unit of energy generated.

A study commissioned by the Office of Science and Innovation in the UK found that research on all of the Generation IV systems faces several key challenges that will require considerable expense and ingenuity to overcome.[70] The report identified significant gaps in materials technology, especially in designing materials that can resist irradiation and neutron damage while operating at high temperatures and minimizing stress corrosion cracking. Fast reactor systems will likely use fuels containing significant quantities of transuranium elements, necessitating a shift away from uranium assemblies to ones based on nitride or carbide fuels; the manufacturing processes for these fuels, however, have not yet even been established. Current modeling and simulation programs are insufficient to map the potential scenarios involving the higher actinides expected to be produced by Generation IV reactors. The report also noted that proliferation resistance has only been demonstrated under laboratory conditions, and that it is unclear how Generation IV technology will be deployed on a larger scale while avoiding dangerous scenarios for nuclear fuel diversion to unstable governments or extra-governmental regimes. Also unknown is how fuels containing high quantities of minor actinides and possibly long-lived fission products will behave, how to design the proper shielding facilities for such substances, and whether conventional waste storage facilities can even handle these unconventional waste streams.

Similarly, researchers at North Carolina State University concluded that the materials used in conventional reactors will not be suitable for

Generation IV technology.[71] Zirconium alloys, for instance, are currently used as fuel cladding in light and heavy water reactors, but will not work under the higher-temperature environments envisioned by Generation IV proponents. Other core components made from low-alloy ferric steels, such as pressure valves, will no longer suffice; in fact, the pressure vessels needed to handle expected temperatures from Generation IV reactors would likely double the size of existing reactors. The researchers noted that a lack of fast spectrum irradiation facilities and high-temperature testing facilities greatly restricts the ability of scientists and engineers to even design, test, and evaluate the necessary structural materials for advanced reactors.

In short, the Generation IV strategy relies on inventing new materials for individual components before the reactor design itself can even be tested. This may explain why some analysts have called Generation IV reactors "paper reactors" — technologies that have yet to be built and exist only on paper.[72]

Researchers for the European Commission agreed and stated in 2007 that an unexpected technological breakthrough must occur before Generation IV technology can become feasible, stating that they found it "inconceivable that the long term objective of sustainable development of nuclear fission energy" could be met with existing technology.[73] Similarly, Exelon invested a 12.5% share in a Generation III+ reactor project in South Africa only to bail out a few years later, citing the project's astronomical costs as the main concern. Another independent international task force studying the feasibility of Generation IV reactors argued that the technical and financial risks seem too difficult to overcome.[74]

To conclude, when the different costs of nuclear power are added together, building a new nuclear power plant is by far the single most expensive source of electricity on the market. As Table 5 depicts, the marginal levelized cost of electricity for a 1,000-MWe facility built in 2009 would be 41.2–80.3 cents/kWh, presuming one actually takes into account construction, operation and fuel, reprocessing, waste storage, and decommissioning. This estimate, moreover, is conservative as it excludes things difficult to quantify or disaggregate, such as the damages from nuclear lifecycle-related greenhouse gas emissions, subsidies, security costs, and research costs. Projections for waste storage and decommissioning also

Table 5: Marginal Levelized Cost for Building a New 1,000-MWe Nuclear Power Plant (cents/kWh, in 2009 USD)

Type of Expense	Cost (cents/kWh)
Construction and operating costs (including fuel costs)	18–30
Reprocessing costs	0.7–1.1
Waste storage costs	20–45
Decommissioning costs	2.5–4.2
Total	41.2–80.3

Note: Construction and operating costs include fuel costs. Security costs and research costs are excluded because no quantifiable estimates exist as of yet.

represent the low end of existing estimates, focus only on reactors (rather than enrichment facilities or uranium mills), and assume storage for only 10,000 years; in actuality, such costs could be ten times greater. In aggregate, the price tag for a new nuclear facility is four to eight times the marginal levelized cost for other power sources on the market, including both fossil fuels and renewable sources of electricity. Nuclear power is ludicrously expensive, so much so that no rational investor or utility executive can afford it if their accounting of costs is full (inclusive of all elements of the nuclear fuel cycle) and fair (taking into account the thousands of future years needed for waste storage).

Endnotes

[1] Stephen Salsbury, "Facing the Collapse of the Washington Public Power Supply System," in Todd La Porte (ed.), *Social Responses to Large Technical Systems: Control or Anticipation* (Dordrecht, Netherlands: Kluwer, 1991), pp. 61–97.

[2] Edward Vine, Marty Kushler, and Dan York, "Energy Myth Ten — Energy Efficiency Measures Are Unreliable, Unpredictable, and Unenforceable," in Benjamin K. Sovacool and Marilyn A. Brown (eds.), *Energy and American Society — Thirteen Myths* (Dordrecht: Springer, 2007), p. 267.

[3] Amory B. Lovins, Imran Sheikh, and Alex Markevich, "Forget Nuclear," *Rocky Mountain Institute Solutions* 24(1) (Spring, 2008), p. 23.

[4] Benjamin K. Sovacool and Christopher Cooper, "Nuclear Nonsense: Why Nuclear Power Is No Answer to Climate Change and the World's Post-Kyoto

Energy Challenges," *William & Mary Environmental Law and Policy Review* 33(1) (Fall, 2008), pp. 1–119.

[5] L. Echávarri, "Is Nuclear Energy at a Turning Point?," *Electricity Journal* 20(9) (2007), pp. 89–97.

[6] Tom Doggett, "NRC Expects Requests for 7 New Nuclear Reactors," Reuters News Service, March 19, 2009.

[7] Patrik Soderholm, "Fuel Choice in West European Power Generation Since the 1960s," *OPEC Review* 22(3) (September, 1998), p. 213.

[8] Greenpeace, *France's Nuclear Failures: The Great Illusion of Nuclear Energy* (Amsterdam: Greenpeace International, November 2008).

[9] Stephen Thomas, Peter Bradford, Anthony Froggatt, and David Milborrow, *The Economics of Nuclear Power* (London: Greenpeace International, 2007).

[10] Source: M.V. Ramana, "Nuclear Power: Economic, Safety, Health, and Environmental Issues of Near-Term Technologies," *Annual Review of Environment and Resources* 34 (2009), pp. 127–152.

[11] Arnulf Grubler, *An Assessment of the Costs of the French Nuclear PWR Program, 1970–2000* (Vienna: International Institute for Applied Systems Analysis, October 2009, IR-09-376).

[12] Ramana (2009).

[13] US Congressional Budget Office, *Nuclear Power's Role in Generating Electricity* (Washington, D.C.: CBO, May 2008).

[14] See Michael T. Hatch, "Nuclear Power and Postindustrial Politics in the West," in John Byrne and Steven M. Hoffman (eds.), *Governing the Atom: The Politics of Risk* (London: Transaction Publishers, 1996), p. 238; and Ralph C. Cavanagh, "Least Cost Planning Imperatives for Electric Utilities and Their Regulators," *Harvard Environmental Law Review* 10 (1986), p. 302.

[15] See Samuel McCracken, "Shoreham and the Environmental Guerrillas," *National Review* (June 24, 1988), p. 14.

[16] Mark Winfield *et al.*, *Nuclear Power in Canada: An Examination of Risks, Impacts and Sustainability* (Toronto: Pembina Institute, 2006), p. 4.

[17] Thomas *et al.* (2007), p. 9.

[18] John Byrne and Steven M. Hoffman, "The Ideology of Progress and the Globalisation of Nuclear Power," in John Byrne and Steven M. Hoffman (eds.), *Governing the Atom: The Politics of Risk* (London: Transaction Publishers, 1996), pp. 11–46.

[19] See Sovacool and Cooper (2008).

[20] Gabrielle Hecht, *The Radiance of France: Nuclear Power and National Identity After World War II* (Cambridge, MA: MIT Press, 1998), p. 2.

[21] Sovacool and Cooper (2008).

[22] Bent Flyvbjerg, Nils Bruzelius, and Werner Rothengatter, *Megaprojects and Risk: An Anatomy of Ambition* (Cambridge: Cambridge University Press, 2003).

[23] Steve Thomas, *The Economics of Nuclear Power: Analysis of Recent Studies* (London: PSIRU, 2005), p. 12.

[24] David Schlissel and Bruce Biewald, *Nuclear Power Plant Construction Costs* (Cambridge, MA: Synapse Energy Economics, 2008), p. 2.

[25] The Keystone Center, *Nuclear Power Joint Fact-Finding* (2007), p. 34.

[26] Joseph Romm, *The Self-Limiting Future of Nuclear Power* (Washington, D.C.: Center for American Progress Action Fund, June 2008), p. 7.

[27] Source: Benjamin K. Sovacool, "Exploring the Hypothetical Limits to a Nuclear and Renewable Electricity Future," *International Journal of Energy Research* 34 (November, 2010), pp. 1183–1194.

[28] Steven Mufson, "Nuclear Projects Face Financial Obstacles," *Washington Post*, March 2, 2010, p. A1.

[29] See Joseph Romm, "Nuclear Bombshell: $26 Billion Cost Killed Ontario Nuclear Bid," ClimateProgress.org, July 15, 2009; and Trevor Findlay, *The Future of Nuclear Energy to 2030 and Its Implications for Safety, Security, and Nonproliferation* (Waterloo, Ontario: Centre for International Governance Innovation, 2010).

[30] Craig A. Severance, *Business Risks and Costs of New Nuclear Power* (Washington, D.C.: Climate Progress, 2009), p. 13.

[31] Jim Harding, "Economics of Nuclear Power and Proliferation Risks in a Carbon-Constrained World," *Electricity Journal* (December, 2007), pp. 66–68. Converted to 2007 USD.

[32] Nathan E. Hultman, Jonathan G. Koomey, and Daniel M. Kammen, "What History Can Teach Us About the Future Costs of U.S. Nuclear Power," *Environmental Science & Technology* (April 1, 2007), pp. 2088–2093.

[33] Romm (2008), p. 4.

[34] Travis Madsen, Johanna Neumann, and Emily Rusch, *The High Cost of Nuclear Power: Why America Should Choose a Clean Energy Future over New Nuclear Reactors* (Baltimore: Maryland PIRG Foundation, March 2009).

[35] Mark Cooper, *The Economics of Nuclear Reactors: Renaissance or Relapse?* (Montpelier, VT: Institute for Energy and the Environment, June 2009).

[36] Madsen *et al.* (2009).

[37] Mark Cooper, *All Risk, No Reward for Taxpayers and Ratepayers: The Economics of Subsidizing the Nuclear Renaissance* (Montpelier, VT: Vermont Law School, November 2009).

[38] See Jungmin Kang, "Analysis of Nuclear Proliferation Resistance," *Progress in Nuclear Energy* 47 (2005), p. 673.

[39] US Department of Energy, *Plutonium Recovery from Spent Fuel Reprocessing by Nuclear Fuel Services at West Valley, New York from 1966 to 1972* (1996), p. 1.

[40] Harvey Wasserman and Norman Solomon, *Killing Our Own: The Disaster of America's Experience with Atomic Radiation* (New York: Delacorte Press, 1982), pp. 133–134.

[41] John L. Campbell, "The State and the Nuclear Waste Crisis: An Institutional Analysis of Policy Constraints," *Social Problems* 34 (1987), p. 24.

[42] Dan Watkiss, "The Middle Ages of Our Energy Policy — Will the Renaissance be Nuclear?," *Electric Light & Power* (May, 2008).

[43] Sovacool and Cooper (2008).

[44] US Department of Energy, *A Roadmap for Developing Accelerator Transmutation of Waste (ATW) Technology: A Report to Congress* (1999), pp. 72–73.

[45] National Research Council, *Review of DOE's Nuclear Energy Research and Development Program* (2008), p. 5.

[46] Peter R. Orszag, Director, Congressional Budget Office, "Costs of Reprocessing Versus Directly Disposing of Spent Nuclear Fuel," Statement before the Senate Committee on Energy and Natural Resources, 110th Congress (2007).

[47] *Ibid.*

[48] Aude Le Dars and Christine Loaec, "Economic Comparison of Long-Term Nuclear Fuel Cycle Management Scenarios: The Influence of the Discount Rate," *Energy Policy* 35 (2006), pp. 2999–3000.

[49] Office of Civilian Radioactive Waste Management, US Department of Energy, *Fiscal Year 2007 Civilian Radioactive Waste Management Fee Adequacy Assessment Report* (2008), p. 2.

[50] US Government Accountability Office, *Nuclear Waste Management: Key Attributes, Challenges, and Costs for the Yucca Mountain Repository and Two Potential Alternatives* (Washington, D.C.: US GAO, November 2009, GAO-10-48).

[51] Sovacool and Cooper (2008).

52 Jason Hardin, "Tipping the Scales: Why Congress and the President Should Create a Federal Interim Storage Facility for High-Level Radioactive Waste," *Journal of Land Resources and Environmental Law* 19 (1999), pp. 293–323.

53 Allison Macfarlane, "The Problem of Used Nuclear Fuel: Lessons for Interim Solutions from a Comparative Cost Analysis," *Energy Policy* 29 (2001), pp. 1379–1389.

54 Sovacool and Cooper (2008).

55 See United Kingdom Atomic Energy Authority, *Decommissioning Fact Sheet* (2005), p. 2; Anibal Taboas, A. Alan Moghissi, and Thomas S. LaGuardia, *The Decommissioning Handbook* (New York: ASME Press, 2004), p. 140; and Nuclear Energy Agency, *The Regulatory Challenges of Decommissioning Nuclear Reactors* (2003), pp. 15–18.

56 United States Nuclear Regulatory Commission, *Decommissioning Nuclear Power Plants* (2008), p. 49. According to the NRC, the 13 US plants in the process of decommissioning as of 2008 are: Dresden Nuclear Power Station, Unit 1; GE VBWR (Vallecitos); Humboldt Bay Power Plant, Unit 3; Fermi 1 Power Plant; Indian Point Unit 1; La Crosse Boiling Water Reactor; Millstone Nuclear Power Station, Unit 1; N.S. Savannah; Peach Bottom Unit 1; Rancho Seco Nuclear Generating Station; San Onofre Nuclear Generating Station, Unit 1; Three Mile Island Nuclear Station, Unit 2; and Zion Nuclear Power Station, Units 1 and 2.

57 World Energy Council, *The Role of Nuclear Power in Europe* (London: WEC, January 2007).

58 Sovacool and Cooper (2008).

59 See Committee on Decontamination and Decommissioning of Uranium Enrichment Facilities, National Research Council, *Affordable Cleanup?: Opportunities for Cost Reduction in the Decontamination and Decommissioning of the Nation's Uranium Enrichment Facilities* (1996), pp. 49–50; and Sovacool and Cooper (2008).

60 US General Accounting Office, *Report to Congressional Committees — Uranium Enrichment: Decontamination and Decommissioning Fund Is Insufficient to Cover Cleanup Costs* (2004), p. 1.

61 Sovacool and Cooper (2008).

62 Greenpeace (2008), p. 15.

63 Michael Krockenberger, "Unclean, Unsafe, and Unwanted: The Nuclear Industry Nightmare," *Habitat* (June, 1996).

[64] Sovacool and Cooper (2008).

[65] NEA and IAEA, *Projected Costs of Generating Electricity: 2005 Update* (Paris: OECD, 2005), pp. 43–45.

[66] International Atomic Energy Agency, *World Distribution of Uranium Deposits (UDEPO) with Uranium Deposit Classification: 2009 Edition* (Vienna: IAEA, October 2009, IAEA-TECDOC-1629).

[67] Stephen I. Schwartz and Deepti Choubey, "The Cost of Nuclear Security," *Los Angeles Times* (January 12, 2009).

[68] Oleg Bukharin, "Security of Fissile Materials in Russia," *Annual Review of Energy and Environment* 21 (1996), pp. 467–496.

[69] National Research Council (2008), p. 5.

[70] Tim Abram and Sue Ion, *Generation-IV Nuclear Power* (2006), p. 3.

[71] K.L. Murty and I. Charit, "Structural Materials for Gen-IV Nuclear Reactors: Challenges and Opportunities," *Journal of Nuclear Materials* 383 (2008), pp. 189–195.

[72] Sovacool and Cooper (2008).

[73] D. Haas and D.J. Hamilton, "Fuel Cycle Strategies and Plutonium Management in Europe," *Progress in Nuclear Energy* 49 (2007), p. 575.

[74] Lori A. Burkhart, "Changing the Fuel Mix: Time for a Nuclear Rescue?," *Public Utilities Fortnightly* (September, 2002), pp. 16–19.

5

Environmental Damages:
"Cutting Butter with a Chainsaw"

On the late summer evening of September 1, 1984, operators at the St. Lucie nuclear power plant in Florida, United States, observed something peculiar. The waters around the power plant's cooling vents, used to suck ocean water into the facility so that it could be condensed into steam, were unusually frothy. As the plant manager sent one of his workers to manually inspect the facility's intake screens, alarms began to ring in unison. The worker quickly reported back that a flotilla of jellyfish was "attacking" the power plant. Hundreds of thousands of moon jellyfish and sea nettles perished in the two-day assault, clogging the cooling system and forcing both reactors to shut down. Stacy Shaw, one of the operators of the facility, told the *New York Times* that "we had to shut down because we couldn't keep the flow of water that we need to run the plant."[1] Plant officials had to rig an elaborate "jellyfish trap" to keep the thousands of creatures floating in from Vero Beach and Stuart out of the cooling system.[2] Then, on September 8, the jellyfish departed and normal operations resumed.

What is striking about this example is that it is an instance where the natural environment — in this case, scores of jellyfish — was threatened by a nuclear power plant and decided to attack it. Most of the time, as this chapter shows, it is the other way around, with nuclear power plants and their affiliated infrastructure inducing four general types of environmental insults: land use impacts, water use and contamination, climate change, and medical and health risks. Underground, open-pit, and *in situ* leaching uranium mining can contaminate water, and have resulted in scores of accidents and environmental pollution in dozens of countries around the world. Nuclear waste storage, at both permanent and temporary storage sites, creates grave environmental concerns. Plant construction and operation have serious impacts on water availability and quality. Reactor vessels are so heavy that rivers may need to be dredged to get them where they need to go, and nuclear cooling systems use more water than any other electricity source, creating a variety of environmental impacts including thermal discharges, impingement, and entrainment, to say nothing of how accidents and spills can disperse tritium and other carcinogenic compounds into water supplies. In terms of climate change, the nuclear fuel cycle is energy-intensive, meaning every part of it has its own affiliated greenhouse gas emissions. In addition, the carbon footprint of nuclear facilities will only get worse as high-grade uranium ores are used up and plants get older, so much so that a typical reactor will be about as "clean" as fossil fuels within 30–40 years. The chapter finally mentions medical and health risks associated with nuclear power, including higher rates of cancer, birth abnormalities, and the presence of radioactive compounds such as strontium-90 found in the teeth of children living near nuclear power plants.

What makes nuclear power so bad for the environment? Nuclear fission produces some of the most hazardous elements on earth, and it also relies on brute force — controlling a nuclear reaction, the same one released in a weapon — instead of grace or properly scaled systems to generate electricity. The nuclear fuel cycle expends enormous amounts of energy to mine, leach, and enrich uranium from the earth, transport it, process it into fuel, place fuel assemblies into reactors, remove them for interim storage, and permanently sequester nuclear waste. The system is equivalent to "cutting butter with a chainsaw — inelegant, expensive, messy, and dangerous."[3]

Land Use

The deleterious impacts on land from the nuclear fuel cycle can be primarily divided into uranium mining and waste storage.

Uranium Mining

Uranium mining is water- and volume-intensive, since quantities of uranium are mostly prevalent at very low concentrations.[4] Uranium is mined in three different ways: underground mining, open-pit mining, and *in situ* leaching. Underground mining extracts uranium much like other minerals, such as copper, gold, and silver, and involves digging narrow shafts deep into the earth.[5] Open-pit mining, the most prevalent type, is similar to strip mining for coal, where upper layers of rock are removed so that machines can extract uranium. Open-pit mining ceased in the US in 1992 due to concerns about environmental contamination and the quality of uranium, as most ore there resides in lower-grade sandstone deposits.[6] Uranium miners perform *in situ* leaching by pumping liquids into the areas surrounding uranium deposits.[7] These liquids include acid or alkaline solutions to weaken the calcium or sandstone surrounding uranium ore.[8] Operators then pump the uranium up into recovery wells at the surface, where it is collected.[9] *In situ* leaching is more cost-effective than underground mining because it avoids the significant expense of excavating underground sites and often takes less time to implement.[10] Nonetheless, it uses significantly more water — as much as seven to eight gallons for every kilowatt-hour of nuclear power eventually generated.[11] Table 1 shows the top global uranium producers from 2002 to 2008, while Table 2 shows the top uranium mining companies. Canada, Kazakhstan, and Australia account for more than half of global production.

The process of uranium mining itself is very wasteful, regardless of the technique. To produce the 25 tons of uranium needed to keep a typical reactor fissioning atoms for one year, 500,000 tons of waste rock and 100,000 tons of mill tailings — toxic for hundreds of thousands of years — will be created, along with an extra 144 tons of solid waste and 1,343 m^3 of liquid waste.[14] Underground mining presents a "significant danger," since the radionuclides uranium-235, radium-226, radon, and strontium-21 accumulate in the soil and silts around uranium mines, often inhaled by miners in the form of radioactive dust.[15] Open-pit mining is prone to sudden emissions of radioactive gases and the degradation of land, as kilometer-wide craters are formed around uranium deposits, which interfere with the flow of groundwater as far as 10 km

Table 1:   **Global Production of Uranium, 2002–2008 (metric tons)**[12]

Country	2002	2003	2004	2005	2006	2007	2008
Canada	11,604	10,457	11,597	11,628	9,862	9,476	9,000
Kazakhstan	2,800	3,300	3,719	4,357	5,279	6,637	8,521
Australia	6,854	7,572	8,982	9,516	7,593	8,611	8,430
Namibia	2,333	2,036	3,038	3,147	3,067	2,879	4,366
Russia	2,900	3,150	3,200	3,431	3,262	3,413	3,521
Niger	3,075	3,143	3,282	3,093	3,434	3,153	3,032
Uzbekistan	1,860	1,598	2,016	2,300	2,260	2,320	2,338
USA	919	779	878	1,039	1,672	1,654	1,430
Ukraine	800	800	800	800	800	846	800
China	730	750	750	750	750	712	769
South Africa	824	758	755	674	534	539	655
Brazil	270	310	300	110	190	299	330
India	230	230	230	230	177	270	271
Czech Republic	465	452	412	408	359	306	263
Romania	90	90	90	90	90	77	77
Germany	221	104	77	94	65	41	0
Pakistan	38	45	45	45	45	45	45
France	20	0	7	7	5	4	5
World	36,033	35,574	40,178	41,719	39,444	41,282	43,853
Tons of U_3O_8	42,529	41,944	47,382	49,199	46,516	48,683	51,716
Percentage of World Demand				65%	63%	64%	68%

Table 2:   **Top Uranium Mining Companies, 2008**[13]

Company	Tons U	%
Rio Tinto	7,975	18
Cameco	6,659	15
Areva	6,318	14
Kazatomprom	5,328	12
ARMZ	3,688	8
BHP Billiton	3,344	8
Navoi	2,338	5
Uranium One	1,107	3
Paladin	917	2
GA/Heathgate	636	1
Other	5,543	13
Total	43,853	100

away.[16] All three types of uranium mines have been shown to release harmful rates of gamma radiation. At five separate mines in Australia — Nabarlek, Rum Jungle, Hunter's Hill, Rockhole, and Moline — gamma radiation levels exceeded safety standards in some cases by 50%, leading to "chronic" exposure to miners and workers.[17]

As is probably obvious to the reader by now, such mining produces a variety of negative environmental impacts. The most direct is occupational hazards. For instance, uranium miners are often exposed to excessively high levels of radon, and hundreds have died of lung cancer and thousands more had their lives shortened. According to reports by the International Commission on Radiological Protection, work-related deaths for uranium mining amount to 5,500–37,500 deaths per million workers per year, compared to 110 deaths for general manufacturing and 164 deaths for the construction industry.[18] Even more worrying is the evidence that there may be no "safe" level of exposure to the radionuclides at uranium mines. One longitudinal medical study found that low doses of radiation, spread over a number of years, are just as "dangerous" as acute exposure.[19]

A second hazard relates to the radioactive waste mines create. To supply even a fraction of the power stations the industry expects to be online worldwide in 2020 would mean generating millions of metric tons of toxic radioactive tailings every single year. These tailings contain uranium, thorium, radium, and polonium, and emit radon-222.[20] Quite simply, uranium mining results in "the unavoidable radioactive contamination of the environment by solid, liquid, and gaseous wastes."[21] A look at the history of uranium mining in 12 countries is most revealing, and troubling.

In Australia, the third-largest producer of uranium in 2008, a detailed investigation of the environmental impacts from the Rum Jungle mine found that it has discharged acidic liquid wastes directly into creeks that flow into the Finniss River and has also gradually eroded the lowlands adjacent to the creeks. Land has been contaminated with radium-226, and "accounting for the radium has been extremely poor with very little focus on radium uptake in the environment or current levels leaching from the site."[22] The Roxby Downs mine has polluted the Arabunna people's traditional land with 80 million tons of annual dumped tailings, in addition to the mine's daily extraction of 30 million liters of water from the Great Artesian Basin. The Ranger mine has seen 120 documented leaks, spills, and breaches of its tailings waste, which has seeped into waterways and contaminated the Kakuda wetlands. The Beverley mine has been fined for dumping liquid radioactive waste into groundwater.[23] The Olympic Dam mine, a vast open-pit mine, has generated windstorms carrying radioactive dust.[24] It also draws 15 million liters of water per day from the Great Artesian Basin, and has dumped five billion liters of toxic and acidic water from tailings into water sources.[25] It may thus come as no surprise that the independent Senate References and Legislation Committee, part of the Australian federal government, documented a pattern at uranium mines where "short-term considerations have been given greater weight than the potential for permanent damage to the environment."[26] In order to maximize production, environmental concerns at Australian uranium mines have been placed second to profits.

In the US, one of the countries with the longest history of uranium mining, mill tailings were discharged with impunity into water sources for most of the 1940s and 1950s. The radium leached from these tailings contaminated thousands of miles of the Colorado River system.[27] Another case occurred between 1966 and 1971, when thousands of homes and commercial buildings in the Colorado Plateau region were found to contain anomalously high concentrations of radon after having been built on uranium tailings taken from piles under the authority of the Atomic Energy Commission (AEC).[28] Wastes from uranium mines in New Mexico have polluted the water supplies of Crownpoint, Coyote Canyon, Mariano Lake, and Smith Lake, and the Diné people of the Navajo Nation living there have discovered aquifers containing more than 200 times the level of uranium considered safe by the World Health Organization (WHO). At a single Navajo reservation, more than 1,000 open mining pits still sit filled with radioactive slurry containing uranium, radium, arsenic, selenium, molybdenum, and other carcinogenic and toxic substances. Children are known to fall into such pits, and houses have been unknowingly built from actual piles of uranium tailings. The Yakama and Spokane reservations in Washington have found radioactive isotopes unique to spent fuel rods in fish caught along the Columbia River.[29] Another study found that, from 1967 until 1986, uranium mine dewatering managed to spread dissolved selenium and molybdenum into the Puerco River in Arizona such that it contaminated 65 km of land with high levels of alpha and beta radiation.[30]

To get a sense for how extremely lethal uranium mining is in the US, consider the case of the Shiprock facility in New Mexico. Of the 150 miners working at the mine, 38 have since died of radiation-induced cancer and another 95 have unusual serious respiratory ailments and cancers (meaning 89% of miners, on aggregate, displayed chronic illnesses). That facility, once closed, left 70 acres of raw untreated tailings almost as radioactive as the ore itself. Other studies have shown higher rates of miscarriages, cleft palates, and birth defects among communities living near uranium mines, to say nothing of the psychological damage and guilt miners feel for infecting their families and loved ones with radioactive particles and illnesses. One study recently argued that uranium mining creates "a health crisis of epidemic proportions" when done near communities.[31] The Jackpile mine in Laguna Pueblo, New Mexico, polluted most of the groundwater for the village of Paguate and spread "heavy contamination" throughout the entire Southwestern US. A local community center, the Jackpile Housing Project, and the tribal council headquarters were all unwittingly built with radioactive materials from the mine. Roads to the mine at Paguate were even repaired with low-grade uranium ore to cut down on asphalt costs. The miners that lived in these communities, as well as their families, also suffered highly elevated cases of lung cancer — rates six times higher than those predicted for ordinary uranium mining. Far from being an anomaly, 52 other mines spread across the canyons and mesas of New Mexico have discharged thousands of tons of tailings directly into rivers and streams.[32] As one environmentalist recently lamented, the result is that the once-pristine Southwest is now home to radioactive peach trees, plutonium-contaminated chilies, radioactive catfish in Cochiti Lake, and tritium-contaminated honeybees.[33]

In Russia, another country with a legacy of mining, the milling and processing of uranium at Streltsovsk, Krasnokamensk, and Bambakai has discharged radioactive pollutants into local water sources and seen tailings seep into water tables. Indoor radon levels within both the mines and nearby homes are "dangerously high," and the new mine at Khiagdinskii no longer bothers to monitor radiation exposure to workers and residents at all.[34]

In Kazakhstan, currently the world's second-largest producer of uranium, uranium mines have contaminated water wells and seeped millions of tons of radioactive sediment into the Koshkar-Ata Lake, which as it dries exposes residents of adjacent villages to radioactive dust.[35] Tajikistan's Leninabad region continues to suffer from high radiation levels caused by the Soviet-era uranium ore mining. Even though mining was halted in 1991, improper disposal of tailings and barely covered storage sites have resulted in radiation levels that are several times higher than internationally accepted standards. Uzbekistan and Kyrgyzstan are similarly threatened, as users of transboundary waters tainted by the chemicals in Tajik tailings, and as home to some of the 23 waste dump sites scattered across Central Asia's Ferghana Valley. With low public awareness about radiation or the harmful effects of these sites, villagers have unknowingly allowed livestock to freely graze and children to play in hazardous areas.

In Brazil, uranium mining and milling facilities have released radionuclides and toxic metals into surface waters along Poços de Caldas. One study of the environmental performance of uranium mines in Brazil found that tailing effluents and radioactive sulfates had seeped into local waterways, and that acid mine and waste rock drainage had spread radon-226, uranium-238, and other dangerous compounds into water supplies.[36]

In China, the country's largest uranium mine, No. 792, is reputed to dump untreated radioactive water directly into the Bailong River, a tributary of the Yangtze River.[37] In India, researchers from the Bhabha Atomic Research Centre in Mumbai found that underground uranium mines at Bhatin, Narwapahar, and Turamdih, along with the uranium enrichment plant at Jaduguda, have discharged mine water and mill tailings contaminated with radionuclides (such as radon) as well as residual uranium, radium, and other pollutants directly into local water supplies. The researchers noted that, since the quality of Indian uranium ore is relatively low, about 99% of the ore processed in mills emerges as waste and tailings.[38]

In South Africa, the Center for Nonproliferation Studies has documented that uranium miners inhale radon gas and radioactive ore dust well above recommended dose limits, and that uranium mines have contaminated water supplies with polluted run-off from mining dumps, seepages from tailings dams, and the discharge of untreated water. Streams around Johannesburg have been measured to contain uranium, sulfates, cyanide, and arsenic from uranium mines. Between 1968 and 1982, millions of tons of mine and mill wastes were generated at just four sites, and

30 billion gallons of improperly treated mine water were discharged into local arroyos and streams. A consequence has been contaminated livestock and abnormally high rates of cancer at some villages.[39]

In other developing countries and emerging economies, the impacts from uranium mining can be even more severe, since such governments often lack strong institutional capacity to enforce environmental regulations and statutes. In Africa, for example, the legacy of uranium mining is terrible health, water contamination, and egregious levels of pollution.[40] Uranium mining also raises serious questions about equity and indigenous people, as 70% of uranium deposits throughout the world are located on indigenous people's lands.[41]

Waste Storage

As Chapter 2 noted, the world's nuclear fleet creates about 10,000 metric tons of high-level spent nuclear fuel each year. About 85% of this waste is *not* reprocessed, and most of it is stored onsite in special facilities at nuclear power plants. Proponents of nuclear power are fond of pointing out that 1 kg of uranium can produce 50,000 kWh of electricity, whereas 1 kg of coal can only produce 3 kWh of electricity. Put another way, the energy released by 1 g of uranium-235 that undergoes fission is equal to 2.5 million times the energy released by burning 1 g of coal. What they do not tell you is that, because nothing is burned or oxidized during the fission process, nuclear plants convert almost all of their fuel to waste with little reduction in mass.

Both commercial fuel cycles are very wasteful. In the once-through cycle, used predominately by the US, Sweden, and Finland, fuel is burned in reactors and not reused, meaning that about 95% of it is wasted. In the closed-loop fuel cycle, utilized by Belgium, France, Germany, the Netherlands, Spain, and the UK, plutonium is extracted from spent fuel, recycled, and reprocessed, but 94% of the fuel is still wasted.[42]

Nuclear power plants therefore have at least five waste streams that contaminate and degrade land:

- They create spent nuclear fuel at the reactor site;
- They produce tailings at uranium mines and mills;

- They routinely release small amounts of radioactive isotopes during operation;
- They can catastrophically release large quantities of pollution during accidents; and
- They create plutonium waste.

Even reprocessing creates waste. For example, France, which reprocesses spent fuel to separate fissile material (pure waste) from usable plutonium, has contributed 1,710 m³ of high-level waste globally — a number that is expected to jump to 3,600 m³ by 2020.[43] Each 1,000-MW reactor, regardless of its fuel cycle, has about 15 billion curies of radioactivity, which is equivalent to the total amount of natural radiation found in all of the oceans.[44] As Figure 1 shows, it will take at least 10,000 years before high-level nuclear waste will reach levels of radiation considered safe for human exposure.

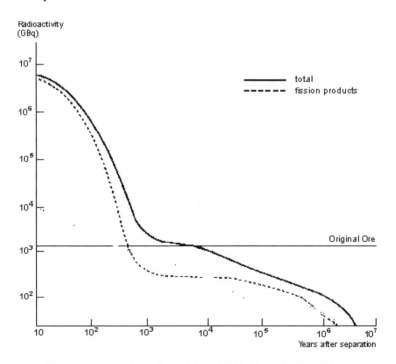

Figure 1:   Decay in Radioactivity of High-Level Nuclear Waste

Note: The straight line shows the radioactivity of the corresponding amount of uranium ore.

The amount of land needed to store nuclear waste can therefore become significant. In 2008, about 57,000 metric tons of uranium existed in the spent fuel inventory from US plants, as well as defense high-level waste. About 85% of this waste was placed 6 m deep in boric acid storage pools at reactor sites while the rest was loaded into 690 dry casks at 42 additional sites, bringing the total number to 131 sites in 39 states (depicted in Figure 2). The dry cask portion of the waste stream is expected to double between 2008 and 2012, and the total amount of waste will reach 119,000 tons by 2035.[45]

France, too, is running out of storage space and existing sites will likely be full by 2015. A 1991 law requiring the creation of a geologic storage facility underground was never implemented due to public opposition.[46] A South Korean underground repository for the permanent disposal of spent nuclear fuel will not be ready until 2041, but interim storage pools will likely reach maximum capacity by 2024.[47] The permanent waste repositories in Finland and Sweden have had all research conducted onsite by the companies themselves with no independent

Figure 2:   Current Spent Fuel Storage Installations in the United States, 2009

review; and the bedrock in both sites is believed to be less stable and full of more cracks than originally believed, with new evidence revealing that copper canisters could be corroded at the site within a century.[48]

Storage of nuclear waste faces a number of daunting challenges, articulated best by a comparative study of waste practices in the US and Japan conducted by researchers at Harvard University and the University of Tokyo.[49] The study identified four key problems with existing schemes to store nuclear waste. First, many of the repositories designed to be temporary are turning into permanent ones. Interim storage, as the name implies, is designed to store waste for a defined period of time where humans can directly monitor it. It is not a substitute for a permanent geologic repository, which must last hundreds of thousands of years. Temporary waste sites are not typically designed to handle contingencies such as earthquakes, tornadoes, and plane crashes, and can operate safely only for a short amount of time.

Second, most communities do not want to host a facility — even a temporary one — for storing nuclear waste. They are concerned about their community becoming a *de facto* site for waste for thousands of years, the health and environmental consequences of an accident, and lower property values. As the authors of the study noted, "local opposition has prevented many past proposed interim storage facilities and other nuclear facilities from being successfully established," and "such objections pose the largest obstacle to building adequate storage capacity for spent nuclear fuel." A recent 2010 assessment confirmed this conclusion by noting that "almost six decades after commercial nuclear energy was first generated, not a single government has succeeded in opening a repository for civilian high-level nuclear waste."[50]

Third, as touched upon in Chapter 3, existing waste sites are prone to accidents, fires, and safety risks. In 1996, as one example, after fuel had been loaded into a dry storage cask at Point Beach, Wisconsin, hydrogen inside the cask ignited as it was being welded and blew the three-ton lid off. The Nuclear Regulatory Commission (NRC) had to take repeated actions throughout the 1990s to address defective welds on dry casks that led to cracks and quality assurance problems; helium had leaked into some casks, increasing temperatures and causing accelerated fuel corrosion.

Fourth, as mentioned in Chapter 4, storing waste is expensive. The total undiscounted lifecycle cost for 40 years of dry cask storage for 1,000 tons of spent fuel, the amount generated by a typical reactor, is US$120–$250 million.[51] When extended to include the entire period that waste must be stored (at least 25,000 years), the costs associated with the existing global reactor fleet jump to a breathtaking US$11.3 trillion.

Even if it is perfected, future Generation IV technology will not solve the problem of radioactive waste. The radiotoxicity for the most hazardous forms of spent nuclear fuel will last at least 100,000 years. Partitioning and transmutation are considered theoretical ways of reducing the waste; but even if technically mastered through some sort of breakthrough, their potential is severely limited. Nuclear engineers at the CEA (*Commissariat à l'énergie atomique*) in France have warned that radiotoxicity can only be reduced by a factor of 10 if all plutonium is recycled, and by a factor of 100 if all minor actinides are burned.[52] This means that, at a minimum, spent fuel will remain dangerously radioactive for at least 1,000 to 10,000 years (or ten centuries), presuming a best-case scenario. Also, the technologies needed to attain this level of waste reduction — either fast reactors or accelerator-driven systems — will require technological breakthroughs in separating actinides, reprocessing advanced fuels, and coupling transmutation technologies to existing reactors. As one study concluded, no single country has successfully deployed partitioning and transmutation technologies, and no attempt has been made to pursue serious regional or international cooperation on these efforts.[53]

The nuclear waste issue, although often ignored in industry press releases and sponsored reports, is the proverbial elephant in the room stopping a nuclear renaissance. As one study concluded:

> The management and disposal of irradiated fuel from nuclear power reactors is an issue that burdens all nations that have nuclear power programs. None has implemented a permanent solution to the problem of disposing of high-level nuclear waste, and many are wrestling with solutions to the short-term problem of where to put the spent, or irradiated, fuel as their cooling pools fill.[54]

Until the issue of waste storage is resolved, the future of nuclear power is highly uncertain.

Water Use and Contamination

The nuclear industry's vast appetite for water has serious consequences, both for human consumption and for the environment. Apart from the water-related impacts of uranium mining, discussed above, three other stages of the nuclear fuel cycle — plant construction, plant operation, and nuclear waste storage — consume, withdraw, and contaminate water supplies. As a result of this monumental need for water, most nuclear facilities cannot operate during droughts and in some cases can actually cause water shortages. For instance, in Germany eight nuclear reactors had to be shut down simultaneously on hot summer days in 2009 for various reasons, many related to the overheating of equipment or of rivers. Droughts and extended periods of high temperature can therefore cripple nuclear power generation, and it is often during these times when electricity demand is highest because of air-conditioning and refrigeration loads and diminished hydroelectric capacity. This disconnect has been poignantly felt in European heat waves, such as in 2003 when France had to cut back 6 GW of capacity and several German reactors operated at 40% capacity.[55] A more recent episode occurred in 2007 in the Southwest of the United States, where nuclear plants were shut down due to lack of water.[56]

Plant Construction

The construction of nuclear power plants can have significant water-related needs and impacts. Some of the largest power plant components, such as turbines, boilers, and reactor cooling towers, have special shipping requirements. In Georgia, US, billions of gallons of water had to be released from Lake Lanier to raise water levels on the lower Chattahoochee River so that replacement steam generators could be shipped to the Farley nuclear power plant near Dothan, Alabama.[57] The Army Corps of Engineers even had to design and maintain a shipping channel from Savannah, Georgia, to Augusta, Georgia, so that power plant equipment could be moved on the river.[58] Since maintenance of the deep-water channel ended in 1979 and Lake Lanier is currently running low on water, power plant operators have warned that rivers in some parts of the South would have to be dredged to allow reactor upgrades and construction of new large power plants to occur.

Plant Operation

Nuclear reactors require massive supplies of water to cool reactor cores and spent nuclear fuel rods, and they use the most water compared to all other electricity-generating facilities, including conventional coal and natural gas facilities.[59] Because much of the water used by nuclear plants is turned to steam, substantial amounts are lost to the local water cycle entirely.

Almost all nuclear power plants employ one of two types of cooling cycles in their generation of electricity. Once-through cooling systems withdraw water from a source, circulate it, and return it to the surface body. As their name implies, once-through cooling systems (or "open-loop" systems) only use water once, as it passes through a condenser to absorb heat. Plant operators commonly add chlorine intermittently to control microbes that corrode pipes and materials. Operators may also add several toxic and carcinogenic chemicals such as hexavalent chromium and hydrazine. After it passes through the plant, heated and treated water is then discharged downstream from its point of intake to a receiving body of water. Since such cooling systems release heated water back to the source, they can contribute to evaporative loss by raising the temperature of receiving water bodies.[60]

Recirculating (or "closed-loop") systems withdraw water and then recycle it within the power system rather than discharge it. Recirculating systems, by recycling water, withdraw much less of it but tend to consume more. Since it is being reused, the water requires more chemical treatment to eliminate naturally occurring salts and solids that accumulate as water evaporates. To maintain plant performance, water is frequently discharged from the system at regular intervals into a receiving body of water or collection pond.[61] Plant operators call this water "cooling tower blowdown." Once the plants release blowdown, operators treat fresh water with chlorine and biocides before it enters the cooling cycle. Closed-loop systems rely on greater amounts of water for cleaning and therefore return little water to the original source.[62]

In aggregate, nuclear power plants are the most water-intensive of all types of power plants, as confirmed by Table 3 and Figure 3. One nuclear plant in Georgia withdraws an average of 57 million gallons of water every day from the Altamaha River, and it actually consumes 33 million gallons

Table 3:   Water Intensity of Thermoelectric Power Generators[68]

Fuel	Cooling Process	Withdrawal (gal/kWh)	Consumption (gal/kWh)
Fossil/biomass/waste	Once-through cooling	20–50	~0.30
Fossil/biomass/waste	Closed-loop tower	0.30–0.60	0.30–0.48
Fossil/biomass/waste	Closed-loop pond	0.50–0.60	~0.48
Nuclear	Once-through cooling	25.00–60.00	~0.40
Nuclear	Closed-loop tower	0.50–1.10	0.40–0.72
Nuclear	Closed-loop pond	0.80–1.10	~0.72
Geothermal steam	Closed-loop tower	~2.00	~1.40
Solar trough	Closed-loop tower	0.76–0.92	0.76–0.92
Solar tower	Closed-loop tower	~0.75	~0.75
Natural gas combined cycle	Once-through cooling	7.50–20.00	0.10
Natural gas combined cycle	Closed-loop tower	~0.23	~0.18
Coal gasification (IGCC)	Closed-loop tower	~0.25	~0.20

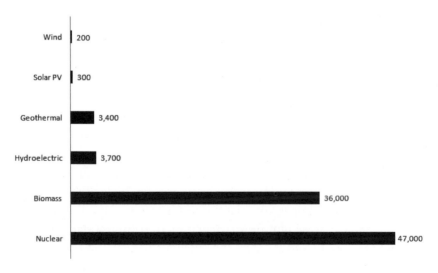

Figure 3:   Water Withdrawn and Consumed by Nuclear and Renewable Power Plants (gallons/MWh)

per day from the local supply (primarily as lost water vapor), which would be enough to service more than 196,000 Georgia homes.[63] The Shearon Harris nuclear reactor, operated by Progress Energy in New Hill, North Carolina (near Raleigh), sucks up 33 million gallons of water a day (and loses 17 million gallons per day due to evaporation). Duke Energy's McGuire plant on Lake Norman, North Carolina, draws in more than 2 billion gallons of water per day.[64] Southern Company's Joseph M. Farley nuclear plant in Dothan, Alabama, consumes about 46 million gallons of water per day (primarily as evaporative loss).[65] In the arid West, where water is scarce, the challenge of cooling nuclear plants is even more daunting. The Palo Verde plant in Arizona is capable of processing 90 million gallons of water for its cooling needs at the plant site each day.[66] Plant operators must purchase treated effluent from seven cities in the Phoenix metropolitan area and had to construct a 35-mile pipeline to carry water from a treatment facility to the plant, which received 22.5 billion gallons of treated effluent in 2000.[67]

At the point of intake, thermoelectric plants bring water into their cooling cycles through specially designed structures. To minimize the entry of debris, water is often drawn through screens.[69] Seals, sea lions, endangered manatees, American crocodiles, sea turtles, fish, larvae, shellfish, and other riparian or marine organisms are frequently killed as they are trapped against the screens in a process known as *impingement.*[70] Organisms small enough to pass through the screens can be swept up in the water flow, where they are subject to mechanical, thermal, and toxic stress in a process known as *entrainment.*[71] Billions of smaller marine organisms, essential to the food web, are sucked into cooling systems and destroyed.[72] Smaller fish, fish larvae, spawn, and a tremendous volume of other marine organisms are frequently pulverized by reactor condenser systems.[73] One study estimated that more than 90% are scalded and discharged back into the water as lifeless sediment that clouds the water around the discharge area, blocking light from the ocean or river floor, which further kills plant and animal life by curtailing the production of oxygen.[74] During periods of low water levels, nuclear plants must extend intake pipes further into rivers and lakes; but as they approach the bottom of the water source, they often suck up sediment, fish, and other debris.[75] Impingement and entrainment consequently account for substantial losses of fish and exact severe environmental consequences.

For example, federal environmental studies of entrainment during the 1980s at five power plants on the Hudson River in New York — Indian Point, Bowline, Roseton, Lovett, and Danskammer — estimated grave year-class reductions in fish populations (the percent of fish killed within a given age class).[76] Authorities noted that power plants were responsible for age reductions as high as 79% for some species; and an updated analysis of entrainment at three of these plants estimated year-class reductions of 20% for striped bass, 25% for bay anchovy, and 43% for Atlantic tomcod.[77] Other researchers evaluated entrainment and impingement impacts at nine facilities along a 500-mile stretch of the Ohio River.[78] The researchers estimated that approximately 11.6 million fish were killed annually through impingement and 24.5 million fish from entrainment. The study calculated economic losses at about US$8.1 million per year.

The US Environmental Protection Agency (EPA) calculated impinge-ment losses from power plants operating near the Delaware Estuary Watershed at more than 9.6 million age 1 equivalents of fish every year, or a loss of 332,000 pounds of fishery yield.[79] The EPA figured that entrainment-related losses were even larger at 616 million fish every year, or a loss of 16 million pounds of catch.[80] Put into monetary value, the recreational fishing losses from impingement and entrainment were estimated to be about US$5 million per year.[81]

Scientists also estimated that the cooling intake systems at the Crystal River Power Plant in Florida, a joint nuclear and coal facility, kill about 23 tons of fish and shellfish every year. As a result, top predators such as gulf flounder and stingray have either disappeared or changed their feed-ing patterns.[82] In other parts of Florida, the economic losses induced from four power plants — Big Bend, P.L. Bartow, F.J. Gannon, and Hookers Point — were estimated to be as high as US$18.1 million.[83] Similarly, in Southern California, marine biologists and ecologists found that the San Onofre nuclear plant impinged nearly 3.5 million fish in 2003.[84]

A less noticed, but still important, impact is that water intake and discharge often alter natural patterns of water levels and flows. Such flows, part of the hydrological cycle, have a natural rhythm that differs daily,

weekly, and seasonally.[85] Plants and animals have adapted to these fluctuations, and such variability is a key component of ecosystem health.[86]
However, withdrawals and discharges of water at nuclear plants alter this
natural variability by withdrawing water during drought conditions or
discharging it at different times of the year, with potentially serious (albeit
not well-understood) consequences to ecosystem and habitat health.

Nuclear power plants also alter the temperatures of lakes, rivers,
and streams.[87] The data on temperature at intake and discharge
points collected by the US Energy Information Administration demonstrated that more than 150 once-through units had summer or winter
discharges with water temperature deltas (large temperature differences between intake and discharge waters) greater than 25°F.[88] In
some cases, the thermal pollution from centralized power plants can
induce eutrophication — a process whereby the warmer temperature
alters the chemical composition of the water, resulting in a rapid
increase of nutrients such as nitrogen and phosphorus.[89] Rather than
improving the ecosystem, such alterations usually promote excessive
plant growth and decay, favoring certain weedy species over others and
severely reducing water quality.[90] In riparian environments, the
enhanced growth of choking vegetation can collapse entire ecosystems.[91] This form of thermal pollution has been known to decrease the
aesthetic and recreational value of rivers, lakes, and estuaries, and
complicate drinking water treatment.[92]

For example, a team of Indian scientists studying heated water
discharges from the Madras Atomic Power Station located at Kalpakkam
in India noted that substantial additions of sodium hypochlorite to
seawater decreased viable counts of bacteria and plankton by 50%
around the reactor site.[93] They also discovered that the plume of thermal
pollution was greater at the power plant's coastal location because the
tidal movements altered its direction and enhanced its magnitude. A
team of Korean marine biologists and scientists utilized satellite thermal
infrared images of the Younggwang nuclear power plant on the west
coast of Korea and found that the plant's thermal pollution plume
extended more than 100 km southward.[94] The researchers documented
that the power plant directly decreased the dissolved oxygen content of
the water, fragmented ecosystem habitats, and reduced fish populations.

Lastly, and most seriously, nuclear power plants create wastewater contaminated with radioactive tritium and other toxic substances that can leak into nearby groundwater sources. In December 2005, for example, Exelon Corporation reported to authorities that its Braidwood reactor in Illinois had, since 1996, released millions of gallons of tritium-contaminated wastewater into the local watershed, prompting the company to distribute bottled water to surrounding communities while local drinking water wells were tested for the pollutant.[95] When caught for its mistake, rather than admit responsibility, Exelon ran a sleek advertising campaign to convince citizens of Illinois that the tritium exposure was "natural" and "can be found in all water sources."[96] The incident led to a lawsuit by the Illinois Attorney General and the State Attorney for Will County, who claimed that "Exelon was well aware that tritium increases the risk of cancer, miscarriages, and birth defects, and yet they made a conscious decision to not notify the public of its risk of exposure."[97]

Similarly, in New York, a faulty drain system at Entergy's Indian Point Nuclear Plant on the Hudson River caused thousands of gallons of radioactive waste to be leaked into underground lakes.[98] The NRC accused Entergy of not properly maintaining two spent fuel pools that leaked tritium and strontium-90, cancer-causing radioactive isotopes, into underground watersheds, with as much as 50 gallons of radioactive waste seeping into water sources per day.[99]

Such examples are not isolated and have not been chosen selectively. As of February 2010, 27 of the 104 reactors operating in the US have been documented leaking radioactive tritium into watersheds.[100] In the UK, the Sellafield reprocessing facility has been accused of contaminating parts of the Irish Sea with radioactive pollutants[101]; and from 1967 to 1969, France dumped more than 12,000 m^3 of high-level waste from the reprocessing plant at Marcoule directly into the ocean.[102]

Nuclear Waste Storage

At reactor sites, even when not generating electricity, nuclear plants must use water continuously — often about 10% of the water needed for normal operation — to cool spent nuclear fuel rods. After the complete shutdown of a nuclear reactor, it continues to produce residual heat that takes days to decay significantly. Nuclear plants need water to remove the decay heat produced by the reactor core, and also to cool the equipment and buildings used to provide the core's heat removal. Service water must lubricate oil coolers for the main turbine and chillers for air-conditioning, in essence cooling the equipment that in turn cools the reactor. These service water needs can be quite high: 52,000 gallons of water are needed per minute in the summer to merely service the Hope Creek plant in New Jersey; 30,000 gallons per minute for the Millstone Unit 2 in Connecticut; and 13,500 gallons per minute for the Pilgrim plant in Massachusetts.[103]

Climate Change

From a climate change standpoint, nuclear power is no improvement over renewable energy resources, despite recent claims by the Nuclear Energy Institute that nuclear power is "clean-air energy."[104] Reprocessing and enriching uranium requires a substantial amount of electricity, often generated from fossil fuel-fired power plants; and uranium milling, uranium mining, uranium leaching, plant construction, and decommissioning all produce substantial amounts of greenhouse gases. As Chapter 2 explained, in order to enrich natural uranium, it is converted to uranium hexafluoride (UF_6) and then diffused through permeable barriers. In 2002, the Paducah uranium enrichment plant in Kentucky released 197.3 metric tons of freon, a greenhouse gas far more potent than carbon dioxide, through leaking pipes and other equipment.[105] Data collected from one uranium enrichment company revealed that it takes a 100-MW power plant running for 550 hours to produce the amount of enriched uranium needed to fuel a 1,000-MW reactor, of the most efficient design currently available, for just one year. According to the *Washington Post*, two of the US' most polluting coal plants, in Ohio and Indiana, produce electricity primarily for uranium enrichment.[106]

When one takes into account the carbon-equivalent emissions associated with the entire nuclear lifecycle, nuclear plants contribute significantly to climate change and will contribute even more as stockpiles of high-grade uranium are depleted. An assessment of 103 lifecycle studies of greenhouse gas-equivalent emissions for nuclear power plants found that the average CO_2 emission over the typical lifetime of a plant was about 66 g for every kilowatt-hour, or the equivalent of some 183 million metric tons of CO_2, in 2005.[107] The specific numbers from this study are presented in Table 4. If the global nuclear industry were taxed at a rate of US$24 per ton for the carbon-equivalent emissions associated with its lifecycle, the cost of nuclear power would increase by about US$4.4 billion per year.[108] A second, follow-up, peer-reviewed study found that the best-performing reactors had associated lifecycle emissions of 8–58 gCO_2/kWh, but that other reactors emitted more than 110 gCO_2/kWh.[109] A secondary impact is that, by producing large amounts of heat, nuclear power plants contribute directly to global warming by increasing the temperature of water bodies and micro-climates around each facility.[110]

The carbon-equivalent emissions of the nuclear lifecycle will only get worse (not better) because, over time, reprocessed fuel is depleted, necessitating a shift to fresh ore, and reactors must utilize lower-quality ores as higher-quality ones are depleted. Table 5 illustrates this clearly: with lower-grade uranium ore, the emissions profile from nuclear power plants almost doubles from 66 gCO_2e/kWh to over 112 gCO_2e/kWh. The Oxford Research Group projected that, because of this inevitable shift to lower-quality uranium ore, if the percentage of world nuclear capacity remains what it is today, by 2050 nuclear power would generate as much carbon dioxide per kilowatt-hour as comparable natural gas-fired power stations.[111] This bears repeating: at current levels of electricity generation, by 2050 nuclear plants will be producing as much greenhouse gas as some fossil fuel plants. With very low ore grades in use, some nuclear power plants currently emit the equivalent of 337 gCO_2/kWh, making them *already* close to the equivalent emissions from gas-fired power plants.[112]

For these reasons, an integrated sustainability analysis conducted in Australia found that nuclear plants are poor substitutes for other less greenhouse gas-intensive generators. The analysis demonstrated that wind turbines have one-third the carbon-equivalent emissions of nuclear power over their lifecycle; and hydroelectric turbines, one-fourth the carbon-equivalent emissions.[113] A separate study from *Nature* found that nuclear power plants emit two times more equivalent greenhouse gases than solar

Table 4:   Lifecycle Greenhouse Gas Emission Estimates for Nuclear Power Plants

Location	Assumptions	Fuel Cycle	Individual Estimate (gCO_2e/ kWh)	Total Estimate (gCO_2e/ kWh)
Canada	CANDU heavy water reactor, 40-year lifecycle, high-quality natural uranium ore, enriched and charged with fossil fuel generators	Front end	0.68	15.41
		Construction	2.22	
		Operation	11.9	
		Back end	—	
		Decommissioning	0.61	
United Kingdom	35-year lifecycle, average load factor of 85%, uranium ore grade of 0.15%	Front end	56	84–122
		Construction	11.5	
		Operation	—	
		Back end	—	
		Decommissioning	16.5–54.5	
Switzerland	100-year lifecycle, Gösgen pressurized water reactor and Liebstadt boiling water reactor	Front end	3.5–10.2	5–12
		Construction	1.1–1.3	
		Operation	—	
		Back end	0.4–0.5	
		Decommissioning	—	
Switzerland, France, and Germany	40-year lifecycle, existing boiling water reactors and pressurized water reactors using UCTE nuclear fuel chains	Front end	6–12	7.6–14.3
		Construction	1.0–1.3	
		Operation	—	
		Back end	0.6 and 1.0	
		Decommissioning	—	
China	20-year lifecycle, once-through nuclear cycle using centrifuge technology	Front end	7.4–77.4	9–80
		Construction	1.0–1.4	
		Operation	—	
		Back end	0.6–1.2	
		Decommissioning	—	
United Kingdom	Analysis of emissions for construction of Sizewell B pressurized water reactor	Front end	—	11.5
		Construction	11.5	
		Operation	—	
		Back end	—	
		Decommissioning	—	

(*Continued*)

Table 4: (*Continued*)

Location	Assumptions	Fuel Cycle	Individual Estimate (gCO$_2$e/ kWh)	Total Estimate (gCO$_2$e/ kWh)
Germany	Analysis of emissions for a typical 1,250-MW German reactor	Front end	20	64
		Construction	11	
		Operation	—	
		Back end	33	
		Decommissioning	—	
United States, Europe, and Japan	40-year lifecycle, 85% capacity factor, mix of diffusion and centrifuge enrichment	Front end	12–21.7	16–55
		Construction	0.5–17.7	
		Operation	0.1–10.8	
		Back end	2.1–3.5	
		Decommissioning	1.3	
Japan	Analysis of base-case emissions for operating Japanese nuclear reactors	Front end	17	24.2
		Construction	2.8	
		Operation	3.2	
		Back end	0.8	
		Decommissioning	0.4	
Sweden and Japan	40-year lifecycle for Swedish Forsmark 3 boiling water reactor and 30-year lifecycle for Japanese boiling water reactor, advanced BWR, and fast breeder reactor	Front end	1.19–8.52	2.82–22
		Construction	0.27–4.83	
		Operation	—	
		Back end	1.19–8.52	
		Decommissioning	0.17	
Australia	Analysis of emissions for existing Australian light water reactors with uranium ore of 0.15% grade	Front end	4.5–58.5	10–130
		Construction	1.1–13.5	
		Operation	2.6–34.5	
		Back end	1.7–22.2	
		Decommissioning	0.1–1.3	
Australia	Analysis of emissions for existing Australian heavy water reactors with uranium ore of 0.15% grade	Front end	4.5–54	10–120
		Construction	1.1–12.5	
		Operation	2.6–31.8	
		Back end	1.7–20.5	
		Decommissioning	0.1–1.2	

(*Continued*)

Table 4: (*Continued*)

Location	Assumptions	Fuel Cycle	Individual Estimate (gCO$_2$e/ kWh)	Total Estimate (gCO$_2$e/ kWh)
Egypt	30-year lifecycle for a pressurized water reactor operating at 75% capacity	Front end	23.5	26.4
		Construction	2.0	
		Operation	0.4	
		Back end	0.5	
		Decommissioning	—	
World	Analysis of emissions for existing nuclear reactors	Front end	36	88–134
		Construction	12–35	
		Operation	—	
		Back end	17	
		Decommissioning	23–46	
World	Analysis of emissions for existing nuclear reactors	Front end	39	92–141
		Construction	13–36	
		Operation	—	
		Back end	17	
		Decommissioning	23–49	
World	Analysis of emissions for existing nuclear reactors assuming 0.06% uranium ore, 70% centrifuge and 30% diffusion enrichment, and inclusion of interim and permanent storage and mine land reclamation	Front end	16.26–28.27	112.47–165.72
		Construction	16.8–23.2	
		Operation	24.4	
		Back end	15.51–40.75	
		Decommissioning	39.5–49.1	
Japan	60-year lifecycle, light water reactor reference case, emissions from 1960 to 2000	Front end	5.9–118	10–200
		Construction	1.3–26	
		Operation	2.0–40	
		Back end	0.7–14	
		Decommissioning	0.1–2	

(*Continued*)

Table 4:   (*Continued*)

Location	Assumptions	Fuel Cycle	Individual Estimate (gCO₂e/ kWh)	Total Estimate (gCO₂e/ kWh)
World	Analysis of emissions for construction and decommissioning of existing reactors	Front end	—	3
		Construction	~2	
		Operation	—	
		Back end	—	
		Decommissioning	~1	
United States	40-year lifecycle of 1,000-MW pressurized water reactor operating at 75% capacity factor	Front end	9.5	15
		Construction	1.9	
		Operation	2.2	
		Back end	1.4	
		Decommissioning	0.01	

energy and about seven times more than wind energy.[114] The author's own calculations, using exclusively peer-reviewed scientific literature, suggest that nuclear power plants are worse than every type of renewable energy generator (see Table 6). Further details on the climate benefits of renewable energy and energy efficiency are offered in Chapter 7.

Medical and Health Risks

As a final, health-related disadvantage, normally functioning nuclear reactors are still correlated with higher risks of cancer and unexplained deaths. Put simply, a proliferation of nuclear power plants inevitably means more nuclear workers and more residents exposed to low-level ionizing radiation, with increased health risks attendant to this exposure.[115]

Reactors create more than 100 dangerously radioactive chemicals, including strontium-90, iodine-131, and cesium-137 — the same toxins found in the fallout from nuclear weapons. Some of these contaminants, such as strontium-90, remain radioactive for 600 years; concentrate in the food chain; are tasteless, odorless, and invisible; and have been found in the teeth of babies living near nuclear facilities. Strontium-90 mimics milk as it enters the body and concentrates in bones and lactating breasts to

Table 5:   Emissions for the Nuclear Fuel Cycle Relying on Lower-Grade Uranium Ore

Nuclear Process	Estimate (gCO$_2$e/kWh)
Front End (total)	16.26–28.27
Uranium mining and milling (soft and hard ores) (uranium grade of 0.06%)	10.43
Refining of yellowcake and conversion to UF$_6$	2.42–7.49
Uranium enrichment (70% UC, 30% diff)	2.83–8.03
Fuel fabrication	0.58–2.32
Construction (total)	16.8–23.2
Reactor Operation and Maintenance (total)	24.4
Back End (total)	15.51–40.75
Depleted uranium reconversion	2.10–6.24
Packaging of depleted uranium	0.12–0.37
Packaging of enrichment waste	0.16–0.46
Packaging of operational waste	1.93–3.91
Packaging of decommissioned waste	2.25–3.11
Sequestration of depleted uranium	0.12–0.35
Sequestration of enrichment waste	0.16–0.44
Sequestration of operational waste	1.84–3.73
Sequestration of decommissioned waste	1.98–2.74
Interim storage at reactor	0.58–2.32
Spent fuel conditioning for final disposal	0.35–1.40
Construction, storage, and closure of permanent geologic repository	3.92–15.68
Decommissioning (total)	39.5–49.1
Decommissioning and dismantling	25.2–34.8
Land reclamation of uranium mine (uranium grade of 0.06%)	14.3
Total	112.47–165.72

cause bone cancer, leukemia, and breast cancer. Babies and children are 10–20 times more susceptible to its carcinogenic effects than adults.[116] Plutonium is so dangerous that one pound evenly distributed could cause cancer in every person on earth; also, it remains radioactive for 500,000 years.[117] It enters through the lungs and mimics iron in the body, migrating to bones (where it can induce bone cancer or leukemia) and to the liver

Table 6: Lifecycle Greenhouse Gas Emissions for Renewable, Fossil-Fueled, and Nuclear Sources of Electricity Supply

Technology	Capacity/Configuration/Fuel	Estimate (gCO_2e/kWh)
Wind	2.5 MW, offshore	9
Hydroelectric	3.1 MW, reservoir	10
Wind	1.5 MW, onshore	10
Biogas	Anaerobic digestion	11
Hydroelectric	300 kW, run-of-river	13
Solar thermal	80 MW, parabolic trough	13
Biomass	Forest wood co-combustion with hard coal	14
Biomass	Forest wood steam turbine	22
Biomass	Short rotation forestry co-combustion with hard coal	23
Biomass	Forest wood reciprocating engine	27
Biomass	Waste wood steam turbine	31
Solar photovoltaic	Polycrystalline silicon	32
Biomass	Short rotation forestry steam turbine	35
Geothermal	80 MW, hot dry rock	38
Biomass	Short rotation forestry reciprocating engine	41
Nuclear	Various reactor types	66
Natural gas	Various combined cycle turbines	443
Fuel cell	Hydrogen from gas reforming	664
Diesel	Various generator and turbine types	778
Heavy oil	Various generator and turbine types	778
Coal	Various generator types with scrubbing	960
Coal	Various generator types without scrubbing	1,050

(where it can cause primary liver cancer). It crosses the placenta into the embryo and, like the drug thalidomide, causes gross birth deformities; and it also has a "predilection for the testicles, where it induces genetic mutations in the sperm of humans and other animals that are passed on from generation to generation."[118]

Specific medical and epidemiological studies about nuclear power, radiation, and health are frightening, to say the least. One medical study found that those living within 10 km of the La Hague nuclear reprocessing plant in northwest France had a sevenfold increase in risk to the incidence of childhood leukemia.[119] A similar study found twice as much plutonium in the teeth of children living near the Sellafield nuclear reprocessing plant in the UK than in those further away.[120] Even the accident at Three Mile Island (TMI) is not as benign as it originally appeared. One comprehensive study divided the 10-mile area around TMI into 69 study tracts, and then assigned radiation dose estimates and correlated them with incidences of leukemia, lung cancer, and all other types of cancer. The study found that residents living around TMI had abnormally high rates for all three.[121]

One of the most comprehensive studies to date was conducted by the German Childhood Cancer Registry at the University of Mainz, known as the "Epidemiological Study on Childhood Cancer in the Vicinity of Nuclear Power Plants" (or *Epidemiologische Studie zu Kinderkrebs in der Umgebung von Kernkraftwerken* in German, abbreviated as "KiKK").[122] Researchers there looked at childhood cancers and leukemia in the areas around the country's 16 commercial nuclear power plants, and found a "strong" relationship between rates of cancer and proximity to nuclear facilities, especially for those living within 5 km of a plant. During the study period 1980–2003, children less than five years old living within 5 km of a nuclear power plant were more than twice as likely to develop leukemia compared to children living greater than 5 km away.

The depressing news is that the researchers presented many reasons why their findings are conservative, and underplayed the medical risks from nuclear power. They based their radiation risk model on data from the Japanese victims of Hiroshima and Nagasaki, but these survivors were exposed to a single flash of high-energy gamma rays from the atomic bombs. Most were also full-grown adults. Their model thus focused on external sources of radiation and did not take radioactive fallout into account. By contrast, Germans living close to nuclear power plants are chronically exposed over long periods of time, inhale or ingest radioisotopes such as tritium and carbon-14, and encompass a population of children and fetuses, making them quite unlike the Japanese sample. These nuclear power plants also expose their populations to alpha and beta emissions, in addition to gamma rays.

Another reason such estimates may be conservative is that new medical evidence firmly suggests that there may be no such thing as "safe" exposure to radiation. One massive study of 15 countries that monitored

407,391 workers for external radiation exposure, with a total follow-up of 5.2 million person-years, found that even low doses could trigger high rates of cancer.[123] Put another way, there is no safe threshold at which the human body can tolerate the unnatural levels of radiation produced by nuclear reactors and their components.

One can actually draw from existing studies to loosely quantify the health risk per nuclear reactor. Evidence from the US, home to 104 operating nuclear reactors at 65 sites, has documented elevated rates of leukemia and brain cancers at nuclear power plants. Joseph Mangano and his colleagues from the Radiation and Public Health Project estimated that roughly 18,000 fewer infant deaths and 6,000 fewer childhood cancers would occur over a period of 20 years if all reactors in the US were closed — in other words, each nuclear plant is associated with 175 infant deaths and 58 childhood cancers.[124] Applied globally, the world's existing 432 reactors likely cause 75,600 infant deaths (26 times the 2,900 who died in the terrorist attacks of September 11, 2001) and 25,056 childhood cancers every 20 years.

Endnotes

1   Reuters, "Jellyfish Invasion Slows Nuclear Plant in Florida," *New York Times*, September 2, 1984.

2   Sally D. Swartz, "St. Lucie Nuclear Plant Restarts," *Palm Beach Post*, September 8, 1984, p. 1.

3   Amory Lovins, "A Target Critics Can't Seem to Get in Their Sights," in Hugh Nash (ed.), *The Energy Controversy: Soft Path Questions and Answers* (San Francisco: Friends of the Earth, 1979), p. 26.

4   For an overview of uranium mining and the front end of the nuclear fuel cycle, see Benjamin K. Sovacool, "Valuing the Greenhouse Gas Emissions from Nuclear Power: A Critical Survey," *Energy Policy* 36 (2008), pp. 2941–2943.

5   *Ibid.*

6   See EPA, *Uranium Mining and Extraction Processes in the United States* (2006), pp. 2-4–2-5, available at http://www.epa.gov/radiation/docs/tenorm/402-r-08-005-voli/402-r-08-005-v1-ch2.pdf/. ("Conventional refers to open-pit and underground mining. Open-pit mining is employed for ore deposits that are located at or near the surface, while underground mining is used to extract

ore from deeper deposits or where the size, shape, and orientation of the ore body may permit more cost-effective underground mining. Since the early 1960s, most uranium has been mined on a larger scale than earlier mining efforts, and, until recently, by using conventional mining techniques. Radioactive mine wastes from conventional open-pit and underground mines are considered to be TENORM, whose regulatory responsibility resides with EPA or the states. In recent years, ISL [*in situ* leaching] operations (regulated by the NRC or its Agreement States) in the United States are described further below. Those operations have generally replaced conventional mining because of their minimal surface disturbance and avoidance of associated costs.")

7   Sovacool (2008).

8   *Ibid.*

9   *Ibid.*

10 EPA, *Technologically Enhanced Naturally Occurring Radioactive Materials from Uranium Mining* (vol. 2) (2008), pp. AIII-1–AIII-2.

11 US DOE, *Energy Demands on Water Resources: Report to Congress on the Interdependence of Energy and Water* (Washington, D.C.: DOE, 2006), p. 56.

12 Source: World Nuclear Association.

13 Source: *Ibid.*

14 David Thorpe, "Extracting Disaster," *The Guardian*, December 5, 2008.

15 V.N. Mosinets, "Radioactive Wastes from Uranium Mining Enterprises and Their Environmental Effects," *Atomic Energy* 70(5) (1991), pp. 348–354.

16 V.V. Shatalov, M.I. Fazlullin, R.I. Romashkevich, R.N. Smirnova, and G.M. Adosik, "Ecological Safety of Underground Leaching of Uranium," *Atomic Energy* 91(6) (2001), pp. 1009–1015.

17 G.M. Mudd, "Uranium Mining in Australia: Environmental Impact, Radiation Releases and Rehabilitation," in IAEA (ed.), *Protection of the Environment from Ionizing Radiation: The Development and Application of a System of Radiation Protection for the Environment* (Vienna: International Atomic Energy Agency, 2003), pp. 179–189.

18 Roxby Action Collective and Friends of the Earth, *Uranium Mining: How It Affects You* (Sydney: Friends of the Earth, 2004).

19 *Ibid.*

20 Thorpe (2008).

21 Mosinets (1991), p. 348.

22 Mudd (2003).

23 Friends of the Earth, *Uranium Mining for Belgian Nuclear Power Stations: Environmental and Human Rights Impacts* (Brussels: Friends of the Earth, 2008).

24 Thorpe (2008).

25 Michael Krockenberger, "Unclean, Unsafe, and Unwanted: The Nuclear Industry Nightmare," *Habitat* (June, 1996).

26 Thorpe (2008).

27 Earl Cook, "The Role of History in the Acceptance of Nuclear Power," *Social Science Quarterly* 63 (1982), p. 10.

28 *Ibid.*, pp. 3–15.

29 "Bush Administration Pushes for Uranium Mining," *Indian Country Today*, February 6, 2002, p. 18.

30 Peter C. Van Metre and John R. Gray, "Effects of Uranium Mining Discharges on Water Quality in the Puerco River Basin, Arizona and New Mexico," *Hydrological Services Journal* 37(5) (1992), pp. 463–480.

31 Barbara Rose Johnston, Susan E. Dawson, and Gary E. Madsen, "Uranium Mining and Milling: Navajo Experiences in the American Southwest," in Laura Nader (ed.), *The Energy Reader* (London: Wiley-Blackwell, 2010), pp. 132–146.

32 Winona La Duke, "Red Land and Uranium Mining: How the Search for Energy Is Endangering Indian Tribal Lands," in Laura Nader (ed.), *The Energy Reader* (London: Wiley-Blackwell, 2010), pp. 105–109.

33 Margaret Amalia Hiesinger, "The House That Uranium Built: Perspectives on the Effects of Exposure on Individuals and Community," in Laura Nader (ed.), *The Energy Reader* (London: Wiley-Blackwell, 2010), pp. 113–131.

34 Friends of the Earth (2008).

35 *Ibid.*

36 Horst M. Fernandas, Lene H.S. Veiga, Mariza R. Franklin, Valeria C.S. Prado, and J. Fernando Taddei, "Environmental Impact Assessment of Uranium Mining and Milling Facilities: A Study Case at the Poços de Caldas Uranium Mining and Milling Site, Brazil," *Journal of Geochemical Exploration* 52(1) (January, 1995), pp. 161–173.

37 Friends of the Earth (2008).

38 R. Tripathi, S. Sahoo, V. Jha, A. Khan, and V. Puranik, "Assessment of Environmental Radioactivity at Uranium Mining, Processing, and Tailings Management Facilities in Jaduguda, India," *Applied Radiation and Isotopes* 66(11) (2008), pp. 1666–1670.

[39] Friends of the Earth (2008).

[40] Thorpe (2008).

[41] Roxby Action Collective and Friends of the Earth (2004).

[42] See Benjamin K. Sovacool and Christopher Cooper, "Nuclear Nonsense: Why Nuclear Power Is No Answer to Climate Change and the World's Post-Kyoto Energy Challenges," *William & Mary Environmental Law and Policy Review* 33(1) (Fall, 2008), pp. 1–119.

[43] Jean Marie Gras *et al.*, "Perspectives on the Closed Fuel Cycle — Implications for High-Level Waste Matrices," *Journal of Nuclear Materials* 362 (2007), p. 385.

[44] "Radioactive Wreck: The Unfolding Disasters of U.S. Irradiated Nuclear Fuel Policies," *Nuclear Monitor* 643 (March 17, 2006).

[45] Barry D. Solomon, "High-Level Radioactive Waste Management in the USA," *Journal of Risk Research* 12 (December, 2009), pp. 1009–1024.

[46] Yves Marignac, Benjamin Dessus, Helene Gassin, and Bernard Laponche, *Nuclear Power: The Great Illusion* (Paris: Global Chance, October 2008).

[47] Chang Min Lee and Kun-Jai Lee, "A Study on Operation Time Periods of Spent Fuel Interim Storage Facilities in South Korea," *Progress in Nuclear Energy* 49 (2007), pp. 323–333.

[48] "Further Nuclear Reactor Construction Delays," *Helsingin Sanomat*, August 11, 2007, p. 11.

[49] Matthew Bunn, John P. Holdren, Allison Macfarlane, Susan E. Pickett, Atsuyuki Suzuki, Tatsujiro Suzuki, and Jennifer Weeks, *Interim Storage of Spent Nuclear Fuel: A Safe, Flexible, and Cost-Effective Near-Term Approach to Spent Fuel Management* (Cambridge, MA and Tokyo: A Joint Report from the Harvard University Project on Managing the Atom and the University of Tokyo Project on Sociotechnics of Nuclear Energy, June 2001).

[50] Trevor Findlay, *The Future of Nuclear Energy to 2030 and Its Implications for Safety, Security, and Nonproliferation* (Waterloo, Ontario: Centre for International Governance Innovation, 2010), p. 18.

[51] Allison Macfarlane, "Interim Storage of Spent Fuel in the United States," *Annual Review of Energy and Environment* 26 (2001), pp. 201–235.

[52] See Massimo Salvatores, "Nuclear Fuel Cycle Strategies Including Partitioning and Transmutation," *Nuclear Engineering and Design* 235 (2005), p. 812.

[53] Gopi Rethinaraj, "Nuclear Safety Issues: Review," Address at the Lee Kuan Yew School of Public Policy, National University of Singapore, April 22, 2008, p. 11.

54   Macfarlane (2001), p. 202.
55   Gala Colover, "Summer 2003: A Lesson for the Future?," *EU Energy* 65 (September 12, 2003); and Reuters, "Heatwave Halves Output at Some German N-Plants," August 5, 2003.
56   Mitch Weiss, "Southern US Drought Could Dry Up Coolant Water and Force Nuclear Plants to Shut Down," Associated Press, January 24, 2008.
57   Rob Pavey, "Reactors May Ride on River," *Augusta Chronicle,* June 5, 2008, p. B1.
58   *Ibid.*
59   See Benjamin K. Sovacool and Kelly E. Sovacool, "Preventing National Electricity-Water Crisis Areas in the United States," *Columbia Journal of Environmental Law* 34(2) (July, 2009), pp. 333–393.
60   *Ibid.*
61   *Ibid.*
62   *Ibid.*
63   See Sara Barczak and Rita Kilpatrick, Southern Alliance for Clean Energy, "Energy Impacts on Georgia's Water Resources" (2007), p. 1, available at http://cms.ce.gatech.edu/gwri/uploads/proceedings/2003/Barczak%20and%20Kilpatrick.PDF/.
64   Mitch Weiss, "Drought Could Force Nuke-Plant Shutdowns," *USA Today,* January 25, 2008, available at http://www.usatoday.com/weather/drought/2008-01-24-drought-power_N.htm/.
65   Southern Alliance for Clean Energy, "Energy Group Urges Planning for Droughts: Avoid Nuclear and Coal Water Hogs," Press Release (October 25, 2007), p. 2, available at http://64.84.47.47/sights/cleanenergy/mediaRoom/docs/SACEdroughtElectricityPR102507.doc/.
66   Pinnacle West Capital Corp., *2006 Corporate Responsibility Report: Water Management,* available at http://www.pinnaclewest.com/main/pnw/AboutUs/commitments/ehs/2006/ehs/water/default.html/ (last visited March 13, 2009).
67   Pinnacle West Capital Corp., *2000 APS Environmental, Health, and Safety Report* (2001), p. 37, available at http://www.pinnaclewest.com/files/ehs/2000/EHS2000_FullReport.pdf/.
68   See US DOE (2006), p. 38.
69   See Ellen Baum, Clear Air Task Force, *Wounded Waters: The Hidden Side of Power Plant Pollution* (2004), available at http://www.catf.us/resources/publications/files/Wounded_Waters.pdf/; and Linda Gunter *et al., Licensed to Kill: How the Nuclear Power Industry Destroys Endangered Marine Wildlife and*

Ocean Habitat to Save Money (2001), available at http://www.nirs.org/reactorwatch/licensedtokill/LicensedtoKill.pdf/.

70 Baum (2004), p. 8.

71 *Ibid.*

72 Gunter *et al.* (2001), p. 8.

73 *Ibid.*, p. 10.

74 *Ibid.*

75 Weiss (January 25, 2008), p. 1.

76 Jeffrey S. Levinton and John R. Waldman, *The Hudson River Estuary* (New York: Cambridge University Press, 2006), pp. 198–199.

77 *Ibid.*

78 See Baum (2004), p. 8.

79 See *ibid.*, p. 4.

80 *Ibid.*

81 *Ibid.*

82 *Ibid.*, p. 6.

83 *Ibid.*

84 David Lochbaum, "Got Water?," Issue Brief, Union of Concerned Scientists (2007), p. 12.

85 Sandra Postel and Brian Richter, *Rivers for Life: Managing Water for People and Nature* (Washington, D.C.: Island Press, 2003), pp. 9–10.

86 See *ibid.*, pp. 9–11.

87 See Baum (2004), pp. 6–9.

88 *Ibid.*, p. 7.

89 See USGS, "Eutrophication" (2008), available at http://toxics.usgs.gov/definitions/eutrophication.html/ (last visited March 13, 2009).

90 See *ibid.*

91 See SEARCA, *Community-Based Inventory and Assessment of Riverine and Riparian Ecosystems in the Northeastern Part of Mt. Malindang, Misamis Occidental* (2005), pp. 15–16, available at http://www.searca.org/brp/pdfs/monographs/River_1st%20gen.pdf/.

92 See Burak Güneralp and Yaman Barlas, "Dynamic Modeling of a Shallow Freshwater Lake for Ecological and Economic Sustainability," *Ecological Modeling* 167 (2003), p. 115.

93 P. Saravanan, A. Priya, B. Sundarakrishnan, P. Venugopalan, T. Subbarao, and S. Jayachandran, "Effects of Thermal Discharge from a Nuclear Power Plant

on Culturable Bacteria at a Tropical Coastal Location in India," *Journal of Thermal Biology* 33(7) (2008), pp. 385–394.

94 Y.-H. Ahn, P. Shanmugam, J. Lee, and Y. Kang, "Application of Satellite Infrared Data Mapping of Thermal Plume Contamination in Coastal Ecosystem of Korea," *Marine Environmental Research* 61 (2006), pp. 186–201.

95 Environment News Service, "Illinois Sues Exelon for Radioactive Tritium Releases Since 1996," March 21, 2006, available at http://www.ens-newswire.com/ens/mar2006/2006-03-21-02.asp/.

96 The utility noted, "[t]ritium is an isotope of hydrogen that produces a weak level of radiation. It is produced naturally in the upper atmosphere when cosmic rays strike atmospheric gases and is produced in larger quantities as a by-product of the nuclear energy industry. When combined with oxygen, tritium has the same chemical properties as water. Tritium can be found at very low levels in nearly all water sources." See Exelon Corporation, "Update: Tritium Remediation Efforts Progressing Well," Press Release (March 8, 2007), available at http://www.exeloncorp.com/NR/rdonlyres/0796AA69-770A-43E8-A60D-892CFF54EDCD/2978/20070308BraidwoodExelonRemediationUpdate1.pdf/.

97 Environment News Service (2006).

98 Abby Luby, "Leaks at Indian Point Created Underwater Lakes," *North County News*, February 28, 2008, available at http://www.abbylu.com/pdfs/ENVIRONMENT/indianpointleaks.pdf/.

99 *Ibid.*

100 Associated Press, "Leaks Spotlight Aging Nuclear Plants," February 1, 2010.

101 Macfarlane (2001).

102 Greenpeace, *France's Nuclear Failures: The Great Illusion of Nuclear Energy* (Amsterdam: Greenpeace International, November 2008).

103 Sovacool and Cooper (2008).

104 *Ibid.*

105 Arjun Makhijani, Lois Chalmers, and Brice Smith, *Uranium Enrichment: Just Plain Facts to Fuel an Informed Debate on Nuclear Proliferation and Nuclear Power* (Takoma Park, MD: Institute for Energy and Environmental Research, 2004), p. 8.

106 Peter Asmus, "Nuclear Dinosaur," *Washington Post*, July 6, 2005, p. A17.

107 Sovacool (2008), pp. 2940–2953.

108 The calculation works like this: In 2005, 435 nuclear plants supplied 16% of the world's power, constituting 368 GW of installed capacity generating

2,768 TWh of electricity. With every TWh of nuclear electricity having carbon-equivalent lifecycle emissions of 66,000 tons of CO_2, these plants emitted a total of some 182.7 million tons. If each ton cost US$24, the grand total would be about US$4.4 billion every year.

[109] Jef Beerten, Erik Laes, Gaston Meskens, and William D'haeseleer, "Greenhouse Gas Emissions in the Nuclear Life Cycle: A Balanced Appraisal," *Energy Policy* 37(12) (December, 2009), pp. 5056–5068.

[110] Bo Nordell and Bruno Gervet, "Global Warming Is Global Energy Storage," Paper No. 454, in *Proceedings of the Global Conference on Global Warming-2008 (GCGW-08)*, 6–10 July 2008, Istanbul, Turkey.

[111] F. Barnaby and J. Kemp, *Secure Energy? Civil Nuclear Power, Security, and Global Warming* (Oxford: Oxford Research Group, 2007).

[112] Beerten *et al.* (2009).

[113] Integrated Sustainability Analysis, *Life-Cycle Energy Balance and Greenhouse Gas Emissions of Nuclear Energy in Australia* (Sydney: University of Sydney, November 3, 2006).

[114] Quirin Schiermeier, Jeff Tollefson, Tony Scully, Alexandra Witze, and Oliver Morton, "Electricity Without Carbon," *Nature* 454 (August, 2008), pp. 816–823.

[115] Richard W. Clapp, "Nuclear Power and Public Health," *Environmental Health Perspectives* 113(11) (2005), pp. 720–721.

[116] Helen Caldicott, "Nuclear Power Isn't Clean, It's Dangerous," *Sydney Morning Herald*, August 27, 2001.

[117] Helen Caldicott, *Nuclear Madness* (New York: W. W. Norton & Co., 1994).

[118] Caldicott (2001).

[119] A.V. Guizard, O. Boutou, D. Pottier, X. Troussard, D. Pheby, G. Launoy, R. Slama, and A. Spira, "The Incidence of Childhood Leukemia Around the La Hague Nuclear Waste Reprocessing Plant," *Journal of Epidemiology and Community Health* 55 (2001), pp. 469–474.

[120] Mitchell O'Donnell, N.D. Priest, L. Strange, and A. Fox, "Variations in the Concentration of Plutonium, Strontium-90, and Total Alpha-Emitters in Human Teeth Collected in the British Isles," *Science of the Total Environment* 201 (1997), pp. 235–243.

[121] S. Wing, D. Richardson, D. Armstrong, and D. Crawford-Brown, "A Reevaluation of Cancer Incidence Near the Three Mile Island Nuclear Plant: The Collision of Evidence and Assumptions," *Environmental Health Perspectives* 105 (1997), pp. 52–57.

122 Rudi Nussbaum, "Childhood Leukemia and Cancers Near German Nuclear Reactors: Significance, Context, and Ramifications of Recent Studies," *Journal of Occupational Environmental Health* 15(3) (2009), pp. 318–323.

123 E. Cardis, M. Vrijheid, M. Blettner, E. Gilbert, M. Hakama, C. Hill *et al.*, "Risk of Cancer After Low Doses of Ionizing Radiation: Retrospective Cohort Study in 15 Countries," *British Medical Journal* 331(77) (2005), pp. 7508–7518.

124 Benjamin K. Sovacool, "Is the Nuclear Option the Right One for Malaysia?," *New Straits Times* (Malaysia), August 12, 2009, p. 17.

6

Political and Social Concerns: "Broken Plowshare"

At the peak of the nuclear era, the US federal government initiated Project Plowshare, a program using nuclear weapons for "peaceful nuclear explosions." The Project was named directly from the Bible itself, specifically Micah 4:3, which states that God will beat swords into plowshares, and spears into pruning hooks, so that no country could lift up weapons against another. By 1961, the first detonation of the program occurred with Project Gnome, which exploded a 10-kiloton device in a salt dome to study isotopes near Carlsbad, New Mexico, followed by 26 other blasts over 11 years costing taxpayers more than US$770 million.

Proposed uses included building sea-level canals into deserts; widening the Panama Canal; constructing a new shipping lane through Nicaragua named the Pan-Atomic Canal; cutting pathways through mountains for highways; connecting inland river systems; creating underground aquifers in Arizona; and even exploding five nuclear weapons to produce a harbor in Cape Thompson, Alaska, along with a channel from the harbor to the ocean (aptly termed Project Chariot). Project Carryall, proposed by the Atomic Energy Commission (AEC) in 1963, would have detonated 22 nuclear weapons to excavate a massive road through the Bristol Mountains in the Mojave Desert so that the California Division of Highways could have an alternate route to Interstate 40 and the Santa Fe Railway could be extended.[1]

The negative impacts from Project Plowshare's 27 nuclear explosions, however, were numerous. Project Gnome vented radioactive steam over the very press gathering that was called to confirm its safety. The next blast, a 104-kiloton detonation in Yucca Flat, Nevada, displaced 12 million tons of soil and resulted in a radioactive dust cloud that rose 12,000 feet and plumed towards the Mississippi River. Other consequences — blighted land, relocated communities, tritium-contaminated water, radioactivity, and fallout from debris being hurled high into the atmosphere — were ignored and downplayed until the program was terminated in 1977, due in large part to public opposition.[2]

The lessons from Project Plowshare to the current nuclear industry are salient. It shows how initial optimism with a new technology can fade into disillusionment and contempt. It reveals how something intended to improve national security can unwittingly do the opposite if it fails to fully consider social, political, and environmental consequences. It also underscores that public resentment and opposition can stop projects in their tracks. This chapter argues that nuclear reactors have, in fact, become what Gene Rochlin calls "broken plowshares."[3] Rather than demonstrating the peaceful use of atomic energy, and improving collective international security, the chapter discusses a pernicious assortment of security-related problems with modern nuclear power plants, including grid vulnerability, reactor insecurity, the proliferation of weapons of mass destruction, the danger of military conflict, and theft and accidents during transportation of nuclear materials. Moreover, the institutions in charge of nuclear power are decidedly anti-democratic, and tend to erode accountability and good governance in the communities where nuclear power plants operate.

In terms of physical security, nuclear plants face at least three types of security risks. They rely on a brittle and inefficient transmission and distribution network that is prone to accidents, attack, and sabotage; power plants and reactor cores themselves offer tempting targets for terrorists; and the fissile material produced from nuclear reactions can be used to make radioactive weapons of mass destruction, for use by rogue states or terrorist regimes bent on producing the greatest amount of human carnage.

Transmission and Distribution Vulnerability

I shall start with one of the least controversial ways in which nuclear power plants degrade energy security: grid vulnerability. Relying on centralized nuclear plants to transmit and distribute electric power creates unavoidable (and costly) vulnerabilities. Electric transmission and distribution systems, when coupled to nuclear power plants, constitute brittle infrastructure that can be easily disrupted, curtailed, or attacked. During the coal miners' strike of 1976 in Britain, a leader of the power engineers famously remarked that "the miners brought the country to its knees in 8 weeks; we could do it in 8 minutes."[4] Centrally generated power requires an overly complex distribution system, which is subject to cascading failures easily induced by severe weather, human error, sabotage, or even the interference of small animals. "Continuous electrical supply," notes physicist Amory Lovins, "now depends on many large and precise machines, rotating in exact synchrony across half a continent, and strung together by an easily severed network of aerial arteries whose failure is instantly disruptive."[5]

The International Energy Agency (IEA) has also noted that centralized energy facilities create tempting targets for terrorists because they would need to attack only a few, poorly guarded facilities to cause large, catastrophic power outages.[6] Thomas Homer-Dixon, Chair of Peace and Conflict Studies at the University of Toronto, cautioned that it would take merely a few motivated people with minivans, a limited number of mortars, and a few dozen standard balloons to strafe substations, disrupt transmission lines, and cause a "cascade of power failures," costing billions of dollars in direct and indirect damage.[7] Paul Gilman, former Executive Assistant to the US Secretary of Energy, has argued that the time needed to replace affected infrastructure would be "on the order of Iraq, not on the order of a lineman putting things up a pole."[8]

The security issues facing centralized electricity supply are almost as serious as they are invisible. In 1975, the New World Liberation Front bombed assets of the Pacific Gas and Electric Company more than ten times, and members of the Ku Klux Klan and San Joaquin Militia have been convicted of attempting to attack electricity infrastructure. Outside of the US, organized paramilitaries such as the Farabundo Marti National

Liberation Front were able to interrupt more than 90% of electric service in El Salvador and penned manuals for successfully attacking power systems.

Several recent trends in the electric utility industry have increased the vulnerability of transmission and distribution (T&D) infrastructure, and thereby made nuclear generation riskier and less reliable. To improve their operational efficiency, many utilities and system operators have increased their reliance on automation and computerization. Low margins and various competitive priorities have encouraged industry consolidation, with fewer and bigger facilities and intensive use of assets centralized in one geographical area. Power control systems are more centralized, spare parts inventories have been reduced, and subsystems are highly integrated across the entire business. Restructuring and consolidation have resulted in lower investment in security in recent years, as cash-strapped utilities seek to minimize costs and maximize revenue available for other areas.[9]

Plant and Reactor Insecurity

Moving away from the transmission grid, the nuclear industry currently faces a multitude of other security-related problems: failure to withstand bombings and acts of sabotage, poor security training, inability to maintain core integrity in case of an attack involving a large aircraft, and vulnerability to attacks at fuel facilities and research reactors.

Most telling is that scores of actual attacks and attempted attacks on nuclear facilities have occurred over the past five decades, including acts of terrorism, armed incursion, sabotage, and bombings. The International Policy Institute for Counter-Terrorism database lists some 167 terrorist incidents involving a nuclear target for the period 1970–1999. According to a Russian intelligence official, during the years 1995 to 1997, 50 instances of nuclear blackmail occurred within the Russian Federation.[10] While a bit dated, between 1970 and 1980, 174 incidents of sabotage and attacks against energy facilities occurred in the US.[11] Between 1966 and 1977, there were 10 terrorist incidents against European nuclear installations alone; and from 1969 to 1975, there were 240 bomb threats against nuclear facilities in the US (along with 14 actual or attempted bombings).

Table 1 presents a sample of some of the confirmed security incidents that occurred at nuclear power plants from 1961 to 2007.

Adding to the danger is the fact that, once attacks succeed on a nuclear facility, the security devices and shielding infrastructure installed to protect the occupiers from invaders and radioactive releases can actually repel attempts to stop an incursion. Also, reactors in the process of being refueled have their containment structures open, meaning an attack would release far more radioactive material into the surrounding environment.[13] Furthermore, shut-down reactors present their own safety risks because of reduced oversight and management, with some guarded only by a barbed wire fence.[14]

Despite the history of past attacks, many nuclear power plants and operators have been criticized for having poor security protocols and ineffective security measures. In 1999, after failing a mock attack at the Waterford 3 Nuclear Plant in Taft, Louisiana, plant managers argued that the exercise was unfair and asked for a second try. A Navy SEAL team returned one year later to conduct a "more rigorous exercise against the plant" and "just ate them alive."[15] As an article in *Science* noted:

> Today's nuclear plants are vulnerable to common-mode failures, such as station blackout events, that could result in core damage in as little as 2 hours. Terrorists could exploit these weaknesses to maximize the severity of an attack. During a series of security exercises between 1991 and 2001, mock attackers were able to simulate the destruction of enough equipment to cause a meltdown at nearly 50 percent of U.S. nuclear plants.[16]

Follow-up exercises in 2002 found that 37 of 81 nuclear plants tested in the US failed their Operational Safeguards Readiness Evaluation.[17] In Russia, military teams have successfully conducted exercises where they have penetrated nuclear facilities multiple times and obtained nuclear materials. In the UK, the Office for Civil Nuclear Security (OCNS) mentioned two peaceful incursions into the Sizewell B nuclear power station site in Suffolk, intended by Greenpeace to highlight the fallibility of security arrangements. Even after the second incursion, the station operator failed to make the necessary improvements to security, and the next year other

Table 1: **Attacks at Nuclear Power Plants and Facilities, 1961–2007**[12]

Date	Location	Description
January 3, 1961	Idaho National Engineering Laboratory, United States	A worker allegedly uses a radioactive control rod to commit a murder-suicide, causing a criticality accident and three deaths in the process
November 10, 1972	Birmingham, Alabama, United States	Melvin C. Cale, Louis Cale, and Henry D. Jackson, Jr. hijack Southern Airways Flight 49, a Douglas DC-9, and threaten to crash it into the nuclear reactor at Oak Ridge, Tennessee, if they do not receive US$10 million in cash; the plane ends up crash-landing in Cuba
March 3, 1973	Lima, Argentina	Guards at a newly completed nuclear power plant are overpowered by an attack by 15 armed men in a shootout that lasts for two hours
August 15, 1975	Monts d'Arrée Nuclear Power Plant, Brittany, France	Terrorists explode two bombs at the power plant's cooling lake and air chimney in an attempt to cause a meltdown
October 10, 1977	Trojan Nuclear Power Station, Oregon, United States	The Environmental Assault Unit of the New World Liberation Front explodes a bomb next to the facility's visitor center to raise awareness against atomic energy
December 18, 1977	Lemoniz Nuclear Power Station, Spain	Members of the Basque separatist group Euskadi Ta Askatasuna (ETA) attempt to overpower power plant guards in a plan to blow up its reactors; one terrorist is killed and the attack is repelled

(Continued)

Table 1: (*Continued*)

Date	Location	Description
March 17, 1978	Lemoniz Nuclear Power Station, Spain	The ETA tries a second time to destroy the power plant, this time exploding a bomb at one of the steam generators, killing two workers and injuring 14
January 3, 1979	Wilmington Nuclear Fuel Facility, North Carolina, United States	A subcontractor for the General Electric fuel processing plant in Wilmington, North Carolina, steals 100 pounds of uranium dioxide by loading it into the trunk of his car and threatens to disperse it in an unnamed city unless he is paid US$100,000
February 10, 1979	Surry Nuclear Power Station, Virginia, United States	Two saboteurs spray sodium hydroxide on four fuel assemblies in an attempt to induce a meltdown
June 13, 1979	Lemoniz Nuclear Power Station, Spain	The ETA again plants a bomb in one of the turbine rooms, igniting a fuel tank of oil that kills one worker and forces a shutdown of the plant for one year
November 24, 1979	Gösgen Nuclear Power Plant, Switzerland	A bomb explodes at one of the transformers near the reactor a few days after the plant has commenced operation
January 16, 1982	Creys-Malville, France	Four anti-tank rockets are fired at the cooling towers of the Superphénix fast breeder reactor, damaging the containment vessel
December 18, 1982	Koeberg Nuclear Power Station, South Africa	The African National Congress, an anti-apartheid movement, explodes four bombs at the newly commissioned power plant

(*Continued*)

Table 1:   (*Continued*)

Date	Location	Description
June 4, 1985	Bataan Nuclear Power Plant, Philippines	Twenty-six bombs placed by Communist guerrillas are discovered around the facility
December 4, 1995	Blayais Nuclear Power Plant, France	Saboteurs pour salt into one of the cooling systems, attempting to cause a meltdown
December 27, 1997	Crystal River Nuclear Power Plant, Florida, United States	A strike team organized by Donald Beauregard, Commander of the 77th Regiment Militia, is intercepted by security forces outside the plant with explosives stolen from a National Guard Armory along with a 20-mm cannon, a .50-caliber machine gun, and pipe bombs; his plan was to destroy the reactor to cause a blackout in St. Petersburg
September 30, 2000	Bern, Switzerland	An anti-nuclear activist manages to land a motorized parafoil on the roof of a reactor at the Mühleberg Nuclear Power Plant and attempts to trip the reactor before being apprehended by security guards
November 8, 2007	Hartbeespoort Dam, South Africa	Four armed men attempt to penetrate the Pelindaba Nuclear Research Centre to acquire spent fuel rods, and infiltrate the eastern block of the facility before being repelled by a security force; one emergency officer is shot

demonstrators were able to gain access via a fire door to an inner security zone.[18] Papers released in February 2005 under the Freedom of Information Act also revealed that, in the previous year, more than 40 security breaches had occurred at civil nuclear sites throughout the UK.[19]

Even the best-intentioned governments can sometimes make bungling security mistakes. In 2009, for example, the US federal government mistakenly released a "highly confidential" and "sensitive" 266-page report on the Internet detailing the locations of civilian nuclear reactors, weapons laboratories, fuel facilities, and stockpiles of uranium and plutonium. The document provided maps, addresses, pictures, images, and locations of fuel vaults, diagrams of reactors, and updates on the quantity and type of fuel being stored. The document was what one analyst called a "one-stop-shop" for information on US nuclear technology for "thieves or terrorists."[20]

A particularly worrisome vulnerability, given the September 11, 2001, terrorist attacks in the US, is the inability for many existing nuclear plants to maintain core integrity in the face of an aircraft impact. Although designed to handle hurricanes, earthquakes, and other extreme weather events, nuclear reactors have not been built to withstand the impact of a large airliner loaded with fuel.[21] As another study in *Science* warned, "One to two large aircraft are large enough to penetrate a 5-foot-thick reinforced concrete wall of a pressurized water reactor spent fuel storage pool, potentially causing it to be so damaged that it rapidly drains and cannot be refilled from either onsite or offsite resources."[22]

The Swiss Federal Nuclear Safety Inspectorate used dynamic load analysis to realistically assess the structural design margins of reactors in Switzerland to survive crashes involving all types of aircraft of varying weight, fuel load, and speed, and in varying weather conditions.[23] Their comprehensive study found that plants designed in the 1970s were not made to deal with modern aircraft crashes, and that at best plants were designed to withstand the impact of a Boeing 707 with a residual fuel level and an impact velocity of 370 km per hour. Modern aircraft such as Boeing 747-400s or Airbus A380s can carry twice as much fuel and reach impact velocities above 600 km per hour. A follow-up study from the Swiss Federal Nuclear Safety Inspectorate concluded that:

From the construction engineering aspect, nuclear power plants (worldwide) are not protected against the effects of warlike acts or terrorist attacks from the air. . . . [O]ne cannot rule out the possibility that fuel elements in the fuel pool or the primary cooling system would be damaged and this would result in a release of radioactive substances.[24]

While the industry purports that plant structures housing reactor fuel can withstand aircraft impact, multiple reports have cautioned that, for too many plants, the vital control building — the building that, if hit, could lead to a meltdown — is still located outside protective structures and is vulnerable to attack.[25]

Furthermore, when the National Research Council surveyed the safety of American nuclear storage facilities in 2006, they noted that a terrorist attack could induce a zirconium cladding fire, releasing large amounts of hazardous radioactive material. They emphasized that these vulnerabilities could not be eliminated by dry cask storage technologies because newly discharged fuel rods must be stored onsite. As the authors concluded, "successful terrorist attacks on spent fuel pools, though difficult, are possible, and if the attack leads to a zirconium cladding fire, it could result in the release of large amounts of radioactive material."[26] One nuclear engineer confided that "a lot of the spent nuclear fuel casks can be hit with a shoulder-fired missile by someone standing outside the fence."[27] Classified reports have also shown how a well-placed truck bomb could destroy vital plant equipment without having to enter a site's property, and other reports have shown that the water-intake systems at some plants are particularly vulnerable to sabotage by either cutting off the water supply by clogging the intake valve or introducing volatile chemicals into plant cooling systems.[28]

Commercial nuclear reactors are not the only security risks. The Center for International Security and Cooperation at Stanford University has warned that research reactors could also be prone to a terrorist attack. These reactors, which contain highly enriched uranium, were promoted under the US "Atoms for Peace" program of the 1950s to form an "atom bank" of knowledge concerning nuclear energy.[29] Today, about 120 research reactors operate around the world and rely on highly enriched uranium, in addition to 270 shut-down research reactors in 74 countries and an extra 109 that have been decommissioned. Russia houses the most research reactors (62), followed by the US (54), Japan (18), France (15), Germany (14), and China (13). Many small and developing countries also have research reactors, including Bangladesh, Algeria, Colombia, Ghana, Indonesia, Jamaica, Libya, Thailand, and Vietnam. About 20 more reactors are planned or under construction.[30]

Indeed, four reasons suggest that research reactors may be at greater risk of attack than operating commercial reactors. First, research reactors, because of their smaller size, tend to have fewer containment structures. They operate at low power and therefore have weaker safety systems. Most lack adequate exclusion zones to protect against truck bombs, and perimeter protection is often only "a wire fence without anti-vehicle barriers, motion sensors, or electronic/computer based detection and assessment systems one finds at commercial nuclear power plants."[31] They would thus be easier for potential terrorists to infiltrate and attack. Second, because they are smaller, components at research reactors are simpler to disassemble and transport. Research reactor fuel elements can be removed without cranes or industrial equipment by a single person; a typical fuel rod is four feet long and weighs 60 pounds, but is still capable of being used to manufacture a lethal radiological dispersal device.[32] Third, university research reactors tend to be located in or near cities, and in places where many people are constantly going back and forth. About half of the research reactors in the US are within 10 miles of population zones containing half a million people or more, including the metropolitan areas of Washington, D.C., Boston, Massachusetts, Denver, Colorado, and Austin, Texas.[33] This makes them ideally situated for a would-be attack. Fourth, research reactors rarely operate continuously, meaning personnel are present only during the day and some have no staff on hand overnight at all. In particular, "many research reactors operated by universities and sometimes by industry are open to visiting specialists (if not to the general public) and have fewer protective security practices than typical nuclear power plants."[34]

Collectively, these factors mean that research reactors lack the containment needed to mitigate the effects of an attack, have smaller nuclear materials that are more accessible to terrorists, are located near population centers, and have inadequate protection. One government report noted that "sleeping guards, unauthorized access to protected areas, disabled alarms in vital areas, and failure to inspect visitors who set off alarms on metal detectors are all serious security problems" at research reactors.[35]

Weapons Proliferation

The Nobel Prize-winning nuclear physicist Hannes Alfvén once said that "atoms for peace and atoms for war are Siamese twins."[36] Because slightly less than 20 pounds (or 9.07 kg) of plutonium is needed to make a nuclear weapon, every ton of separated plutonium waste has enough material for 110 nuclear weapons. The European Union alone produces 2,500 tons of spent fuel annually, containing about 25 tons of separated plutonium, along with 3.5 tons of minor actinides such as neptunium, americium, and curium as well as 3 tons of long-lived fission products — enough fissile material for 2,750 new nuclear weapons every year. The four countries with the largest reprocessing fleets (Belgium, France, Germany, and the UK) declared more than 190 tons of separated plutonium in 2007, mostly stored in plutonium dioxide powder at above-ground sites and fuel manufacturing complexes — enough for 20,900 nuclear weapons.

Put another way, a typical nuclear reactor produces enough pluto-nium every two months to create a nuclear weapon. Taken as a whole, commercial nuclear reactors already create, every four years, an amount of plutonium equal to the entire global military stockpile. Moreover, the manufacturing of nuclear weapons from spent fuel is not the only risk: 1 kg of plutonium is equivalent to about 22 million kWh of heat energy. A dirty bomb laced with 1 kg of plutonium can therefore produce an explo-sion equal to about 20,000 tons of chemical explosives.[37]

Box 1:   Defining Highly Enriched Uranium and Weapons-Grade Nuclear Material

Highly enriched uranium refers to uranium that has been enriched in the isotope uranium-235 to more than 30%. The critical mass (the bare mass of uncompressed material required to initiate a neu-tron chain reaction) of 50%-enriched uranium is a factor of three greater than that of 93.5%-enriched uranium. Plutonium of any isotopic composition (with the exception of Pu-238) can be used to manufacture a nuclear explosive device; however, weapons design-ers prefer to use plutonium with high levels of Pu-239. Plutonium containing more than 94% Pu-239 is termed "weapons-grade."[38]

Box 2:   Fissile Material and a Typical Nuclear Reactor

A typical pressurized water reactor core contains about 75 tons of uranium. Operators refuel the reactor every 18 months. The fuel elements normally stay in the reactor for three fuel cycles (or 60 months), but the refueling schedule is staggered so that at each refueling the operators take out one-third of the fuel assemblies — the ones that have been in the core for three cycles — and replace them with fresh fuel. Reactor-grade plutonium refers to fuel that has been in the reactor for a full three fuel cycles. Such fuel indeed has a high content of isotopes other than the most desirable plutonium-239. However, a light water reactor operator seeking better plutonium for weapons is not constrained to using the plutonium from irradiated fuel assemblies. For example, if the operator of a newly operating reactor unloaded the entire core after 8 months or so, the contained plutonium would be weapons-grade (with a plutonium-239 content of about 90%). The amount of plutonium produced would be about 2 kg per ton of uranium, or about 150 kg per 8-month cycle. This comes to about 30 bombs' worth. Does a would-be nuclear weapon state need more? If the short refueling cycles were continued, the annual output of weapons-grade plutonium would be about 200 kg (allowing for refueling time), thus illustrating what a standard light water reactor can do when viewed as a plutonium production reactor configured to make nuclear weapons.[39]

There is no shortage of terrorist groups eager to acquire the nuclear waste or fissile material needed to make a crude nuclear device or a dirty bomb.[40] Former Secretary-General of the United Nations Kofi Annan put it best when he stated that "nuclear terrorism is still often treated as science fiction — I wish it were. But unfortunately we live in a world of excess hazardous materials and abundant technological know-how, in which some terrorists clearly state their intention to inflict catastrophic casualties."[41]

The risks are not confined to the reactor site. All stages of the nuclear fuel cycle are vulnerable, including:

- Stealing or otherwise acquiring fissile material at uranium mines;
- Attacking a nuclear power reactor directly;
- Assaulting spent fuel storage facilities;
- Infiltrating plutonium stores or processing facilities;
- Intercepting nuclear materials in transit; and
- Creating a dirty bomb from radioactive tailings.

Actual incidents of nuclear theft and procurement of fissile material are frightening, to put it mildly. They demonstrate both the motive of terrorist groups to acquire such material as well as their capability in procuring it.

From 1993 to 2003, shortly after the collapse of the Soviet Union, authorities documented 917 incidents of nuclear smuggling in Russia, Germany, France, Turkey, Libya, Jordan, and Iran, with some of the more infamous cases presented in Table 2.[42] Based on data available from the International Atomic Energy Agency (IAEA), India has reported 25 cases of missing radioactive materials; of these, 13 have never been recovered and 52% occurred by theft.[43] The Database on Nuclear Smuggling, Theft and Orphan Radiation Sources (DSTO) reported that more than 40 kg of weapons-usable uranium and plutonium have been stolen from poorly protected nuclear facilities in the former Soviet Union during the last decade. Of the 71 countries currently engaged in nuclear activities, only 63 adhere to IAEA safeguards concerning the security of spent fuel and radioactive material.[44] When asked about the Russian military's ability to control its nuclear stockpile, General Alexander Lebed humorously replied that "it's difficult to swim in hydrochloric acid with your legs chopped off."[45] Researchers at the Institute for National Strategic Studies cautioned that nuclear weapons reductions have only exacerbated such a risk, because the warhead destruction process increases the amount of available fissile material and weapons-usable material placed in storage facilities without adequate safeguards.[46]

The explanations behind these incidents are relatively simple: people are fallible, and both good and bad people will invariably work at nuclear facilities. For example, alcoholism and drug abuse are prevalent at Russia's

Table 2:   Cases of Nuclear Theft and Terrorist Acquisition of Fissile Material, 1991–2004

Year	Location	Description
1991–1992	Glazov, Russia	Several kilograms of natural and depleted uranium are stolen from the Chepetsky Metallurgical Plant, but the material is intercepted as it is being smuggled across the border to Poland; 12 people, including plant personnel and security officers, are arrested
1992	Podolsk, Russia	Leonid Smirnov, a chemical engineer, steals 1.5 kg of highly enriched uranium from the Luch Scientific Production Association over the course of five months and attempts to sell it to Middle Eastern suppliers before he is caught by police
1992	Khabarovsk Krai, Russia	Inspectors at the Komsomolsk-na-Amure weapons depot discover that 23 nuclear warheads are unaccounted for
1993	Kola Peninsula, Russia	Servicemen and poorly paid security guards conspire to steal two naval reactor fuel assemblies containing 3.6 kg of enriched uranium from the Andreeva Guba technical base
1993	Izhevsk, Russia	Authorities apprehend a Lithuanian businessperson carrying 60 kg of weapons-grade uranium and a fuel rod in the trunk of his car, believed to be stolen from the Chepetsk Mechanical Plant one year earlier
1993	Murmansk, Russia	Three naval reactor fuel assemblies containing 4.5 kg of enriched uranium are stolen from a fuel storage area at the Sevmorput Shipyard
1995	Amman, Jordan	Agents from the US Central Intelligence Agency seize 100 kg of enriched uranium, gyroscopes, and detonators from a shed at Amman Airport
1995	Moscow, Russia	Chechen terrorists successfully smuggle a dirty bomb composed of C4 plastic explosive and cesium-136 into a heavily populated Moscow park, but the bomb fails to detonate
1998	Rome, Italy	Italian police seize 10 kg of highly enriched uranium and two fuel rods, believed to be stolen from a research reactor in the Congo, in the apartment of an African businessman

(Continued)

Table 2:   (*Continued*)

Year	Location	Description
2001	Bogota, Colombia	Colombian police seize 0.5 kg of weapons-grade plutonium, thought to be smuggled from Iran, in the trunk of a car
2002	Moscow, Russia	General Lebed announces that an inventory of Russia's Special Atomic Demolition Munitions — portable, low-yield nuclear weapons that are the equivalent of the United States' tactical nuclear weapons — revealed that 84 are missing
2004	Prague, Czech Republic	Czech police seize 3 kg of highly enriched uranium from the apartment of a Syrian businessperson
2004	Kiev, Ukraine	Ukrainian authorities seize 6 kg of U-235 from the apartment of two former Russian soldiers

nuclear power plants and reprocessing facilities. One Russian sociologist even commented that "a nuclear power plant does not fight alcoholism, it propagates it. Alcoholics are advantageous for nuclear power plants, they are modest and undemanding."[47] The Ministry for Atomic Energy of the Russian Federation went so far as to say that the nuclear industry there had "a total lack of a culture of security."[48]

In the US, the nuclear industry is known for the weak allegiance of its workers, the high turnover rate (most work at their job for less than six months), poor background checks, inferior equipment, poor communication with local authorities, and little training.[49] Eleven guards at the Trojan Nuclear Power Plant in Oregon were charged with drug dealing, many nuclear power plants have fired guards for being drunk on the job, and one facility even hired a convicted felon who had been imprisoned for (ironically) armed robbery.[50] A carpenter named Carl Drega worked for three separate nuclear power plants despite having an arrest record, and later proceeded to shoot and kill two state troopers, a judge, and a newspaper editor.[51] A computer programmer who worked at the Maine Yankee Nuclear Power Plant left the facility only to kill seven of his coworkers at a small technology company in Massachusetts; he was renowned at the nuclear power plant for sleeping in a coffin and telling colleagues he was "so angry he felt like killing someone."[52] In 1998, an employee at the

Turkey Point facility in Florida had access to critical areas of the plant for more than a month before officials learned of his 14 arrests; and at the Calvert Cliffs facility in Maryland, it took eight months before an employee was discovered to be an illegal immigrant with fake identification papers and an arrest record. Worker sabotage has been reported at no less than eight American reactors: Indian Point (New York), Zion (Illinois), Quad Cities (Illinois), Peach Bottom (Pennsylvania), Fort St. Vrain (Colorado), Trojan (Oregon), Browns Ferry (Alabama), and Beaver Valley (Pennsylvania).[53] Nuclear operators have also been found falsifying reactor records to escape fines and cheating on licensing examinations (including two of the operators who caused the accident at Three Mile Island). As nuclear security officer Richard Kester quipped, "Charles Manson could get access to a nuclear power plant."[54]

To be sure, any sufficiently large industry will have its share of hiring mistakes, poor background checks, sabotage, fraud, falsification, and laziness, whether it involves the manufacturing of Chicken McNuggets or Chevrolets. The only difference between the above incidents and those in other industries is that people at nuclear facilities have access to unique radiological and fissile materials.

Unfortunately, the Nuclear Non-Proliferation Treaty (NPT) and IAEA safeguards appear unable to stop the proliferation of fissile material and the drive to acquire nuclear weapons for at least five reasons. First, the IAEA is thinly staffed, underfunded, and overworked. A recent high-level commission found that the IAEA is already badly overstretched and needs to be considerably strengthened to address the need for monitoring and safeguarding.[55] The agency's 2,000 or so full-time staff have been struggling under a zero-growth budget and are spread thinly across very different mandates. The IAEA must not only devise technical safeguards relating to nuclear power plant safety and perfect tools for assessing the economics and planning of new nuclear reactors, but also oversee the use of nuclear isotopes for medical diagnostics and treatment, manage research programs on new waste management systems, and undertake basic science in experimental nuclear physics. Yet, at a time when the IAEA urgently needs additional resources to carry out its mandates, consensus on the basic bargain that underlies the non-proliferation regime and legitimizes the IAEA is breaking down.[56]

Table 3:   **Global Nuclear Weapons Arsenal, 2010**[58]

Country	Strategic	Non-Strategic	Operational	Total Inventory
Russia	2,600	2,050	4,650	12,000
United States	1,968	500	2,468	9,600
France	300		~300	300
China	180		~180	240
United Kingdom	160		<160	185
Israel	80			80
Pakistan	70–90			70–90
India	60–80			60–80
North Korea	<10			<10
Total	~5,400	~2,550	~7,700	~22,500

Second, the five authorized nuclear weapon states under the NPT have not seriously pursued disarmament.[57] Despite the Cold War being over for almost two decades, more than 22,000 warheads form the combined stockpile of nuclear weapons. Table 3 shows that the US and Russia still have close to 10,000 weapons each, and have shown a reluctance to disarm further.

Third, those states not listed among the five authorized under the NPT to possess nuclear weapons have not paid a significant price for their pursuit of weapons capabilities. States that were never party to the NPT — Israel, India, and Pakistan — have, at various stages after 1968, attained nuclear power status. Among parties to the treaty, North Korea withdrew from the NPT in 2003 and has since embarked on its own ambitious nuclear program, spurring some in neighboring Japan to advocate going nuclear as a deterrent. Meanwhile, Iran, another signatory, has been exercising its right under the NPT to develop "peaceful" nuclear technologies. Even Myanmar is constructing a 10-MW thermal research reactor, with training and equipment from Russia, that could form the basis of a nuclear weapons program.[59]

Fourth, the NPT has been explicitly weakened by a number of bilateral deals made by NPT signatories, notably the US. In 1988, for example, the US agreed to accept leftover spent fuel from the Taiwan Research Reactor, to be stored at a Savannah River facility operated by the Department of

Energy (DOE), in order to build stronger ties and counterbalance Chinese hegemony, even though this violated the NPT. In 1992, under the Swords-to-Plowshares Agreement, the US purchased 500 metric tons of weapons-grade highly enriched uranium from Russia at a cost of US$12 billion, again weakening the NPT. In 1994, the US promised to accept 8,000 spent fuel rods from North Korea as part of an "Agreed Framework" in exchange for light water reactors and a pledge by the North Koreans to abandon their nuclear weapons program.[60] Most recently, after three years of negotiation, the US signed a bilateral nuclear deal — the 123 Agreement — with India in October 2008. The deal allows India to receive nuclear technologies and fuel from the Nuclear Suppliers Group, an international consortium that controls trading in nuclear weaponry, fuel, and technologies; it also exempts India from having to sign onto the NPT.[61] These deals contribute to the notion that a "double standard" exists among NPT signatories, whereby they will flout the norms of the treaty whenever it suits them personally. There is also a general idea that nonproliferation goals are at odds with new research on nuclear power. For members of the G8 to convince other countries not to reprocess spent fuel but then urgently pursue it for themselves is "hypocritical" and "sends the wrong signal" to countries such as Taiwan, Brazil, Turkey, and South Korea, in addition to Iran and North Korea.[62]

Fifth, the NPT says nothing about other aspects of the nuclear fuel cycle, such as uranium mines and mills, from which terrorists could easily acquire fissile material. One recent assessment found that dozens of nations, including uranium producers, remain potential "weak links" in the global defense against nuclear terrorism and tacitly ignore United Nations mandates on controls over fissile material at uranium mines.[63] Niger, a major uranium exporter, and the Democratic Republic of the Congo, the source of uranium for the first atomic bomb, are among the states falling short in complying with UN Security Council Resolution 1540. Uncontrolled freelance mining in the Congo has long worried international authorities, for the raw material for a bomb could fall into the hands of terrorists. Twenty-nine nations have failed to report that they have taken action on nuclear security as required by the 2004 resolution, including the uranium producers Zambia, Malawi, and the Central African Republic.

For these five reasons, a number of high-ranking officials, even within the United Nations, have argued that they can do little to stop states from using nuclear reactors to produce nuclear weapons. A 2009 United Nations report warned that:

> The revival of interest in nuclear power could result in the worldwide dissemination of uranium enrichment and spent fuel reprocessing technologies, which present obvious risks of proliferation as these technologies can produce fissile materials that are directly usable in nuclear weapons. . . . Technical measures alone would not compensate for the limitations of the existing nuclear non-proliferation regime. Some international institutional mechanisms, which are non-technical in nature and involve various political, economic, or diplomatic strategies for controlling access to sensitive materials, facilities or technologies, are needed.[64]

The final sentence of this conclusion is troubling, for it suggests that there is no way to design nuclear systems that are truly proliferation-resistant. Even the Nuclear Energy Agency has acknowledged that the proliferation issue cannot be solved by technology, that any solution must be achieved "primarily through political means."[65] A *Jane's Intelligence Digest* report concluded that a substantial increase in the number of new nuclear power plants worldwide would directly increase the risks associated with nuclear weapons proliferation.[66] "Should a state with a fully developed fuel cycle capability decide for whatever reason to break away from its non-proliferation commitments," commented Mohamed ElBaradei, the former Director of the IAEA, "most experts believe it could produce a nuclear weapon within a matter of months."[67]

The impact of even one single attack involving nuclear or radiological materials could be crippling. (Readers should revisit Chapter 5 for a discussion of the likely economic and human health consequences of a nuclear accident or attack.) The Stimson Center calculated the consequences of nuclear events using a Hazard Prediction and Assessment Capability analysis, and found that "very dangerous levels of radioactivity" require only a small amount of particular substances. For example, as Table 4 shows, 9 g of cobalt-60 — the equivalent of two paper clips' worth

Table 4:   Radioactive Materials That Could Be Used in a Radiological Dispersal Device[70]

Radioactive Material	Half-Life	Specific Activity (Ci/g)	Type of Ionizing Radiation
Cobalt-60	5.3 years	1,100	Beta, high-energy gamma
Cesium-137	30 years	88	Beta, high-energy gamma
Iridium-192	74 days	450–1,000	Beta, high-energy gamma
Strontium-90	29 years	140	Beta, low-energy gamma
Americium-241	433 years	3.4	Alpha, low-energy gamma
Californium-252	2.7 years	536	Alpha, low-energy gamma

of material — could make a dirty bomb capable of mass destruction.[68] Public fear about radiation could further "disrupt social order and over-whelm emergency response and medical systems," adding significantly to economic damages and human casualties.[69]

The ability of the international nonproliferation regime, already weakened by the factors above, to control fissile material may be further undermined by advanced Generation IV nuclear research. A transition to newer reactors could (perhaps oddly) *exacerbate* many of these security risks. The most favored designs for Generation IV reactors tend to improve economics, trying to earn operators greater revenue, but do so at the risk of weaker controls on proliferation and fuel efficiency. Fast neutron and breeder reactors would use plutonium as their primary fuel and, if breeders are ever able to produce more plutonium than they consume, they could produce a virtually inexhaustible supply of weapons-grade material. Reprocessing, a key part of the Generation IV strategy, is also inherently unsafe from a proliferation standpoint. The IAEA has been unable to reduce statistical measurement uncertainties over how much plutonium is produced below 1% for plutonium-uranium extraction (PUREX) reprocessing; this means that there will always be uncertainty about how much pure plutonium is actually produced. For these reasons, reprocessing has been abandoned by scores of other countries. Since 1977, when Japan started operating its Tokaimura pilot plant, no state without nuclear weapons has begun civilian reprocessing. Argentina, Belgium, Brazil, Germany, and Italy shut down their plants, and South Korea and Taiwan abandoned laboratory research. The same problem occurs with thorium. A thorium fuel cycle produces highly radioactive Th-228; and neutron bombardment of thorium produces uranium-233, another fissile material that can be used for nuclear weapons.[71] Consequently, in October 2007, leading arms control experts signed a letter urging the suspension of Generation IV reactor research on the grounds that continuing it could encourage substantial nuclear weapons proliferation.[72] A 2005 study from the Nuclear Transmutation Energy Research Center of Korea also concluded that, despite the rhetoric spouted by Generation IV advocates, no nuclear power systems (encompassing "all kinds" of possible reactors and fuel cycles) will ever be completely proliferation-proof.[73]

Military Conflict

While rarely discussed as a security concern, nuclear reactors become preferred targets during military conflict, and also produce byproducts used to manufacture particularly lethal military weapons.

Over the past three decades, nuclear reactors have been repeatedly attacked during military air strikes, occupations, invasions, and campaigns. To list just a few examples:

- Iran bombed the Al Tuwaitha nuclear complex in Iraq in September 1980;
- An Israeli air strike completely destroyed Iraq's Osirak nuclear research facility in June 1981;
- Iraq bombed Iran's Bushehr nuclear plant (which housed two partly built reactors) six times between March 1984 and 1987;
- The US bombed two small Soviet-built nuclear reactors and the French-supplied Tammuz-2 reactor, as well as uranium hexafluoride conversion and centrifuge enrichment pilot facilities, in Iraq in 1991;
- Iraq launched Scud missiles at Israel's Dimona nuclear power plant in 1991; and
- In September 2007, Israel bombed a Syrian gas-cooled, graphite-moderated nuclear reactor under construction without grid connections in a deserted canyon east of the Euphrates River that would have produced enough plutonium for two bombs within a year of operation.

Perhaps the best of these examples is the US attacks on Iraq in 1991. During its initial wave of precision air strikes in January, the American military targeted energy infrastructure, including three nuclear plants, both to disrupt military systems and to enhance the overall psychological and economic impact of the attacks.[74] Subsequent World Health Organization (WHO) investigations estimated that 31% of animal resources were directly exposed to hazardous radiation and that 42% of arable soil was contaminated; and a United Nations commission visiting Iraq after the Gulf War admitted, "Iraq has, for some time to come, been relegated to a pre-industrial age, but with all the disabilities of post-industrial dependency on intensive use of energy and technology."[75]

Nuclear energy is also unique among all energy sources "in terms of its radiological and chemical hazards and as a potential resource for construction of weapons."[76] The nuclear fuel cycle, especially uranium mining and uranium tails, produces depleted uranium that can be encased in special munitions, which are especially effective against tanks.[77] These radioactive shells pose their own set of security risks. In the early 1970s, the US Army began using depleted uranium metal in "kinetic energy penetrators" designed to puncture tank armor. Thirty-millimeter rounds, used by US forces in the First Gulf War and in Kosovo, can pierce steel armor up to a thickness of 9 cm. Some aircraft, such as the A-10, can fire 3,900 rounds of depleted uranium projectiles a minute. Because less than 10% of the penetrators hit each target, much of the depleted uranium becomes aerosolized and blankets the attack area with radioactivity. Some studies have found that depleted uranium dust can travel for 40 km and remain airborne for a considerable amount of time. Depleted uranium can also be used as protective armor, inserted into a metallic sleeve in the regulator steel plates of a tank.[78]

Many of the world's militaries currently possess depleted uranium weapons, and ammunition containing depleted uranium was used extensively in the 1991 Gulf War, the 1994–1995 conflict in Bosnia-Herzegovina, the 1999 air raids in Kosovo, and the 2003 Gulf War. In the 1991 Gulf War, the US Air Force fired more than 780,000 depleted uranium rounds that ended up discharging 259 tons of uranium, in addition to the 61 further tons fired by Army and Marine forces. Almost 11,000 depleted uranium rounds, corresponding to 3 tons of uranium, were fired around Sarajevo during the North Atlantic Treaty Organization (NATO) air strikes in Bosnia-Herzegovina in 1994 and 1995, and 10 tons of uranium were fired from A-10 planes operating in Kosovo.[79] Shells containing 1,700 tons of depleted uranium were fired during the 2003 Iraq War.[80] Although fiercely contested by military officials, many researchers believe that spent depleted uranium munitions cause cancer, birth defects, and widespread environmental degradation.[81]

Maritime and Transport Security

Nuclear power needs and produces a variety of hazardous materials that must be transported globally over land, sea, and air; these include low- and high-level radioactive waste, spent nuclear fuel, uranium ore, fissile materials, and plutonium obtained from reprocessing. Rail is the most preferred form of land transport, but materials are also sent via road convoys, military aircraft, and specially designed transoceanic vessels. About 20 million parcels of varying weight and size are transported each year.[82] These shipments face a distinct collection of security risks.

In Canada, the uranium needed to create fuel rods must travel more than 4,000 km before the process is complete.[83] In Europe, most uranium is transported 150–805 km by railway, 1,250 km by boat, or 378 km by truck.[84] The long distances involved in the nuclear supply chain are vulnerable to accidents and theft. In April 1988, casks used for transporting spent fuel between Germany, Switzerland, France, and the UK were found to contain unsafe levels of radioactive contamination on the exterior of the casks. The German Environment Minister declared that the industry had been aware of the excessive radiation since the 1980s, but had failed to inform authorities.[85] Seventy accidents involving spent-fuel shipments occurred from 1949 to 1999[86]; and between 1971 and 2001, spent nuclear fuel shipments in the US traversed 1.6 million miles by truck and 120,000 miles by rail and suffered four serious accidents. Applying this accident ratio to future shipments, one study estimated that 160–390 accidents will occur from 2005 to 2040.[87]

Moving away from accidents to terrorism and theft, transportation infrastructure and modes of transport are generally "favorite" targets for terrorists. Figure 1 shows that each year there are hundreds of incidents,

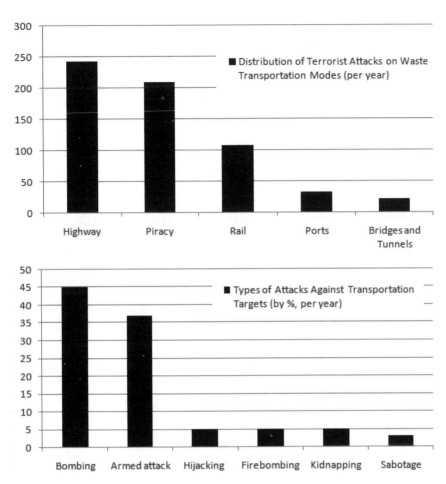

Figure 1:   Distribution and Types of Terrorist Attacks Against Transportation Targets

with about two-thirds (69%) directed against surface transportation targets; the most preferred mode is bombing (45%), followed by armed attack (37%).[88] Nuclear waste shipments are attractive from a tactical viewpoint because they are easy for a small force to seize and because canisters move on known routes and at regular intervals over a long period, with specific known stops, in placarded vehicles, to a single destination. Since storage casks are hot, thermal signatures make them easy to identify using night vision goggles.[89] The September 11, 2001, terrorist attacks, as well as the less well-known 1970 Dawson's Field hijackings (where members of the Popular Front for the Liberation of Palestine hijacked four separate airliners), demonstrate the feasibility of multiple, simultaneous attacks on similar targets. It is conceivable that such a coordinated attack could occur involving nuclear materials in transit.

Land-based shipments are not the only ones at risk. Although it is difficult to pin down exact estimates of shipments of mixed oxide (MOX) plutonium fuel, due to their secrecy, about 80 containers carrying 500 kg of MOX fuel are transported by sea every decade, in addition to 16,500–52,000 other containers of high-level waste shipped annually (including enriched uranium, nuclear waste, contaminated clothing, cleaning residue, chemical resins, and radioactive air filters).[90] Yet, as one legal study concluded, "An attack, wreck, or sinking of a ship carrying nuclear material in a coastal state's waters could have catastrophic effects on their coastal environment and industries; potentially devastating their economy, largely based on coastal resources, and crippling the health and welfare of its people."[91] Similarly, another assessment argued that "long-lived, highly radioactive and radiotoxic nuclear materials could endanger large coastal populations or produce widespread, long-term radioactive contamination of the marine environment," and that "if a vessel carrying such a cargo collided with another vessel causing an intensely hot and long-lasting shipboard fire, then radioactive particles could become airborne, putting all nearby life forms in grave danger of catastrophic health impacts."[92]

The threats that could cause such a scenario can be divided into four types.[93] First, slow-moving vessels carrying nuclear materials could become caught up in sea-based military conflicts of contests over sovereignty. Military conflicts between nations could shut off sea lanes, temporarily or permanently. Examples include a naval conflict over North Korea or the South China Sea, which could clog waterways and shipping lanes, or low-intensity conflicts in Myanmar and East Timor involving sea mines and attacks on ships. Sovereignty disputes could also suddenly intensify into military conflict at sea. More than 30 active territorial disputes involving all major regional powers (China, Japan, South Korea, and India) — and some, such as in the South China Sea, involving six countries (China, Vietnam, Malaysia, Indonesia, Brunei, and the Philippines) — exist in Asia alone. There is also one case in which a Chilean Navy vessel threatened to ram the British-flagged *Pacific Pintail*, carrying nuclear materials, in 1995 as it passed Cape Horn and attempted to enter Chilean waters without authorization.

Second, and more serious, is terrorist threats. Recent advances in telecommunications and commercial logistics have increased the ease with which terrorists can conduct operations at sea. In particular, Southeast Asia has become a "hub" of active maritime terrorism, with the Abu Sayyaf Group, Gerakan Aceh Merdeka, and Jemaah Islamiya all conducting attacks at sea. Al-Qaeda famously attacked the *USS Cole*, while at port in the Gulf of Aden, in 2000 and the *SS Limburg*, anchored off the coast of Yemen, in 2002. The Philippine Moro Islamic Liberation Front also attacked the coastal freighter *Our Lady Mediatrix* in 2000, killing 40 people and wounding 50 more. A few years later, the Abu Sayyaf Group bombed *SuperFerry 14* and killed 116 people in 2004. Terrorist groups are most likely to target seaborne commerce on straits, ships constrained by slow speeds and restricted in the maneuvers they can undertake, as well as vessels encountering navigational hazards or natural chokepoints. Ships containing nuclear fuel and materials fit almost all of these criteria. As Bernard D. Cole, former Captain for the US Navy and Professor at the National War College, put it, "the presence of terrorist groups in maritime Asia is incontrovertible, as unfortunately is the continued presence of the social, economic, and political conditions that give rise to their existence."[94]

Third, and equally serious, is international crime and piracy. Maritime criminals benefit from three types of piracy: hijacking of cargo, opportunistic boarding aimed at robbery or theft, and kidnapping and holding of crew members and officers for ransom. Southeast Asia — where almost all of the nuclear fuel heading for China, Japan, and South Korea passes — is one of the most frequent scenes of piracy, second to Africa. One-quarter of pirate attacks occur in Indonesian waters per year; and the Malacca Strait is especially popular, given its numerous rivers and inlets which make it difficult to secure.[95]

Fourth, companies and captains shipping nuclear materials by sea have been repeatedly unable to keep shipping routes secret and prevent incidences of unauthorized boarding.[96] In 1992, the voyage of the Japanese-flagged *Akatsuki Maru*, carrying 1.7 tons of plutonium from France, was supposed to be kept secret, but public groups found out and prohibited the ship from entering the waters of Argentina, Chile, Malaysia, Portugal, and South Africa. The ship itself was shadowed by a Greenpeace vessel that got so close it was rammed by a Japanese patrol boat. In 1998, Greenpeace activists were able to board *Pacific Swan*, carrying 30 tons of MOX fuel, as it slowly moved through the Panama Canal — enough plutonium to make 60 nuclear weapons; Greenpeace members pulled a small boat alongside *Pacific Swan* under the darkness of morning and boarded the ship with ropes.

There has also been a marked lack of response of shipping states to the sinking of a vessel with nuclear material.[97] Nuclear materials are hazardous to ship by sea, since severe weather and human error become riskier when handling special cargo such as spent fuel. Dangers are exacerbated by long distances, as longer shipping routes require more fuel, increasing the risk and intensity of an explosion on board that could disperse radioactive material over a wide vicinity of sea or land.[98] As one case in point, Figure 2 shows the *MSC Carla*, a Panamanian vessel carrying 11 tons of high-level nuclear waste, which broke in two during rough weather and sank in 1997. Since no state would come to the ship's rescue, its radioactive cargo will gradually release cesium-137 and other radionuclides into the environment off the coast of the Azores as its stainless steel cylinders corrode.[99]

Commentators have argued that existing guidelines are clearly insufficient to handle these sea-based threats. IAEA standards, for example, do not require prior notification to or authorization from transit states during voyages for shipment of nuclear materials, set voluntary (rather than mandatory) recommendations for safeguarding shipments of material, and allow requirements for packing and shipping to be overruled by national security exceptions.[100] The Basel Convention on the Control of Transboundary Movements of Hazardous Wastes and Their Disposal explicitly excludes nuclear waste precisely because of the liability issues involved with an attack or accident.[101] This is why John M. Van Dyke calls nuclear cargos at sea not just dangerous but "ultra-hazardous" goods.[102]

Figure 2:   The *MSC Carla* Sinking off the Coast of the Azores

Note: The *MSC Carla*, which sank off the Azores coast, contained 11 tons of high-level nuclear waste that it will disperse into the marine environment as storage casks degrade underwater.

Community Marginalization

A final challenge relates to democracy and the vitality of communities sur-rounding nuclear facilities. These sociopolitical risks fall into three areas: the marginalization and peripheralization of people and communities liv-ing near nuclear power plants and waste sites, a tendency for nuclear power to go hand-in-hand with authoritarian regimes, and the manipula-tion of public opinion by the nuclear industry to paint a false picture of social acceptance.

First, in terms of marginalization, current nuclear power and waste sites, as well as nominated future sites, express certain characteristics that make them "peripheral communities," set apart from vibrant (and often wealthier) urban communities.[103] These peripheral communities tend to be:

- *Remote,* either geographically separated from population centers or relatively inaccessible;
- *Economically marginal,* with most communities being homogeneous in terms of the social and demographic background and dependent on the nuclear industry as a dominant employer;
- *Politically powerless,* with most key political decisions being made else-where, often in metropolitan centers;
- *Culturally defensive,* with residents expressing ambivalent or ambigu-ous attitudes towards nuclear energy, combined with feelings of isolation and a fatalistic acceptance of nuclear activities; and
- *Environmentally degraded,* meaning residents tend to occupy previ-ously polluted land or are close to places where radioactive risks are already present.

Such factors are reproduced and reinforced in a process known as *periph-eralization.* In essence, nuclear facilities will invariably migrate to communities that lack the political, social, and economic strength to oppose them, especially indigenous peoples and tribes, who are often at the extreme social and geographical periphery of society.[104]

Second, following from the first condition about marginalization, nuclear power systems thrive most in environments that have anti-democratic tendencies. One investigation of nuclear power development in

France, Sweden, and the US found that in each case the demand for reactors had to be created and sustained entirely by the state, was born of military origins and linkages, and relied on governments (rather than private markets) to take a commanding role in nuclear propagation. In each case, early adopters had to be deeply incentivized to participate in the nuclear industry, given its risks and liabilities. Also, all three states prevented public protest from impacting the industry, actively suppressed dissent, discouraged full consideration of alternatives, and built state-sponsored infrastructure for things like waste storage and fuel enrichment.[105]

My own work with Professor Scott Valentine has found the same to be true with nuclear development in France, China, India, Japan, and South Korea.[106] Despite the differences between these five countries, six sociopolitical conditions best explain why nuclear technologies are able to flourish: a strong state involvement in guiding economic development, centralization of national energy planning, campaigns linking technological progress to a national revitalization, influence of technocratic ideology on policy decisions, subordination of challenges to political authority, and low levels of civic activism.

Consider two other examples, one of a more democratic state that is now resisting nuclear power and one of a closed authoritarian state that has continued to embrace it. In Indonesia, under his New Order regime, President Suharto expressed an ambitious desire to facilitate technological supremacy through expensive, large-scale, sophisticated technology. Nuclear energy was introduced to Indonesia in the 1950s through the "Atoms for Peace" program, and the country formed an Institute of Atomic Energy in 1959 and started operating a research reactor in 1961. The vision of a nuclear Indonesia was directly challenged, however, at the turn of the millennium, with parliamentarians arguing that the government should better invest its resources in small-scale energy technologies (like cookstoves and liquefied petroleum gas substitution). Restructuring of the electricity industry placed greater emphasis on keeping electricity prices low, which did not bode well for the capital expense of building a new nuclear power plant. Community leaders (including prominent religious figures) also organized a strong grassroots coalition against nuclear power, which they viewed as too risky and expensive. These threads of resistance coalesced into a formidable political force in 2004 and 2005, exposing the hazards that accompany nuclear energy in media discussions and coordinating social protests. Thus, as of 2009, Indonesia has formally delayed its plans to build its first reactor near Mt. Muria.[107] France, by contrast, is the strongest incarnation of a nuclear state that has resisted democratic input, rejected market liberalization, and maintained centralized control over national discussions of nuclear policy. As one study noted, the "network of actors responsible for development of nuclear energy [in France] has been a closely knit, centralized one, closed to political influence for a long time."[108] Nuclear power has grown significantly in authoritative France, but has stalled in the more democratic Indonesia.

Third, perhaps because the nuclear industry tends not to thrive in places that prioritize the involvement of civil society and that display high levels of civic activism, nuclear advocates have a history of manipulating data and public opinion to serve their own ends. Detailed accident studies carried out by the US Atomic Energy Commission in the 1950s and 1960s (and discussed in Chapter 3) identified the massive consequences of a nuclear meltdown, but were suppressed under pressure from the industry. The results only became public in 1973, after repeated requests made by concerned citizens under the Freedom of Information Act.[109]

More recently, the industry has relied on different tactics to resist or downgrade public participation in licensing and permitting procedures. Some of the more common practices include:

- Setting the timing of hearings so that interveners and the public must gather information and present safety concerns before essential documents are filed by nuclear suppliers and utilities;
- Interpreting rules narrowly to exclude unfavorable evidence and safety concerns; and
- Cutting off public hearings and stopping inquiries into safety issues by prematurely ending proceedings.[110]

These practices are designed to intentionally prevent the public from having a say in where nuclear facilities and infrastructure will go. In other cases, industry representatives display disdain and a complete intolerance of criticism. David J. Rose once commented that raising legitimate concerns about nuclear power in the 1970s and 1980s with members of the nuclear community had "the intelligence, grace, and charity of a duel in the dark with chain saws."[111] In still other cases, the industry response to negative publicity is to flat-out lie. For example, when inspectors recently discovered that 600,000 gallons of boiling radioactive water had turned into steam at the Indian Point Nuclear Power Plant near New York City, and had been released into the lower Hudson Valley for days, operators responded that the escaped radioactive steam was "intentional."[112]

In one of its most intractable problems, that of nuclear waste, the industry has outright manipulated data and limited true public participation to get its way. Some studies of consumer attitudes and public opinions have shown that public groups will support nuclear power expansion if assurances of safe waste disposal are provided, but will not if the waste problem is not resolved.[113] Thus, nuclear power proponents — trade groups, vendors, and utilities — have shifted from a technical discourse, which is full of uncertainty, to a public discourse of inclusive and respectful public consultation about siting as well as criterions of acceptability and safety. Yet, one study of such efforts in Canada found that they do not involve true consultation or participation, whereby citizens have the chance to influence eventual decisions, and are instead public relations exercises used to reinvent the industry.[114] Nuclear groups employ public consultation sessions to (1) demonstrate consent and approval when they do get it, or (2) construct the public as having fragmented values and opinions that will never be overcome when they do not get it, telling regulators they should ultimately defer to the nuclear industry. This situation does not bode well for democracy, the study concluded, as the public is co-opted either way. Public consultation is converted from a means to inform public policy into an end justifying nuclear expansion.

Endnotes

[1] See US Department of Energy, Nevada Operations Office, *United States Nuclear Tests, July 1945 Through September 1992* (Las Vegas, NV: DOE/NV — 209-REV15, December 2000); and US Department of Energy, Nevada Operations Office, Office of Public Affairs and Information, *Project Plowshare* (Washington, D.C.: Government Printing Office, 2001).

[2] See Scott Kirsch, *Proving Grounds: Project Plowshare and the Unrealized Dream of Nuclear Earthmoving* (New Brunswick, NJ: Rutgers University Press, 2005); Frank Kreith and Catherine B. Wrenn, *The Nuclear Impact: A Case Study of the Plowshare Program to Produce Gas by Underground Nuclear Stimulation in the Rocky Mountains* (Boulder, CO: Westview Press, 1976); and Bruce A. Bolt, *Nuclear Explosions and Earthquakes: The Parted Veil* (San Francisco, CA: W.H. Freeman & Company, 1976).

[3] G.I. Rochlin, "Broken Plowshare: System Failure and the Nuclear Power Industry," in J. Summerton (ed.), *Changing Large Technical Systems* (San Francisco: Westview Press, 1994), pp. 231–261.

[4] See Amory Lovins *et al.*, *Small Is Profitable: The Hidden Economic Benefits of Making Electrical Resources the Right Size* (Snowmass, CO: Rocky Mountain Institute, 2002); and Amory B. Lovins and L. Hunter Lovins, *Brittle Power: Energy Strategy for National Security* (Andover, MA: Brick House, 1982), pp. 157–158.

[5] *Ibid.*

[6] International Energy Agency, *Distributed Generation in Liberalized Electricity Markets* (2002), pp. 16–17.

[7] Thomas Homer-Dixon, "The Rise of Complex Terrorism," *Foreign Policy* (January/February, 2002), p. 52.

[8] Quoted in Benjamin K. Sovacool and Christopher Cooper, "Nuclear Nonsense: Why Nuclear Power Is No Answer to Climate Change and the World's Post-Kyoto Energy Challenges," *William & Mary Environmental Law and Policy Review* 33(1) (2008), pp. 1–119.

[9] National Research Council, *Making the Nation Safer: The Role of Science and Technology in Countering Terrorism* (2002), p. 178.

[10] Greenpeace, *Media Briefing: Nuclear Power and Terrorism* (London: Greenpeace United Kingdom, January 2006).

[11] *Ibid.*, p. 83.

[12] Source: Mohammad Saleem Zafar, *Vulnerability of Research Reactors to Attack* (Washington, D.C.: Stimson Center, April 2008); Tilman Ruff, *Nuclear Terrorism* (Australia: Energy Science, November 2006); and William Robert Johnston, "Nuclear Terrorism Incidents" (September 28, 2003).

[13] Daniel J. Barnett, Cindy L. Parker, David W. Blodgett, Rachel K. Wierzba, and Jonathan M. Links, "Understanding Radiologic and Nuclear Terrorism as Public Health Threats: Preparedness and Response Perspectives," *Journal of Nuclear Medicine* 47(10) (2006), pp. 1653–1661.

[14] Jason Hardin, "Tipping the Scales: Why Congress and the President Should Create a Federal Interim Storage Facility for High-Level Radioactive Waste," *Journal of Land Resources and Environmental Law* 19 (1999), pp. 293–323.

[15] Douglas Pasternak, "A Nuclear Nightmare," *U.S. News & World Report,* September 17, 2001.

[16] Edwin S. Lyman, "Revisiting Nuclear Power Plant Safety," *Science* 299 (January 10, 2003), pp. 201–203.

[17] David N. Orrik, Reactor Security Specialist, Nuclear Regulatory Commission, "A Review of Enhanced Security Requirements at NRC Licensed Facilities," Statement before the House Subcommittee on Oversight and Investigations, 108th Congress (April 11, 2002).

[18] OCNS, *The State of Security in the Civil Nuclear Industry and the Effectiveness of Security Regulation, April 2002–March 2003* (2003).

[19] OCNS, *The Organisation and Work of the Office for Civil Nuclear Security* (February, 2005).

[20] William J. Broad, "Secret Nuclear List Accidentally Released," *The New York Times* (June 3, 2009).

[21] Carl Behrens and Mark Holt, "Nuclear Power Plants: Vulnerability to Terrorist Attack," *CRS Report for Congress* (Washington, D.C.: Congressional Research Service, February 4, 2005, RS21131).

[22] Frank N. von Hippel, "Revisiting Nuclear Power Plant Safety," *Science* 299 (January 10, 2003), pp. 201–203.

[23] Swiss Federal Nuclear Safety Inspectorate, *Position of the Swiss Federal Nuclear Safety Inspectorate Regarding the Safety of the Swiss Nuclear Power Plants in the Event of an Intentional Aircraft Crash* (Würenlingen: March 2003, HSK-AN-4626).

[24] Swiss Federal Nuclear Safety Inspectorate (HSK), "Protecting Swiss Nuclear Power Plants Against Airplane Crash," Memorandum (2003), p. 7.

[25] Robert F. Kennedy, Jr., "Nuclear Plants Vulnerable to Attack," *Seattle Post-Intelligencer,* August 5, 2005.

[26] Committee on the Safety and Security of Commercial Spent Nuclear Fuel Storage, *Safety and Security of Commercial Spent Nuclear Fuel Storage* (Washington, D.C.: National Research Council, 2006).

[27] Pasternak (2001).

[28] *Ibid.*

29 George Bunn and Chaim Braun, "Terrorism Potential for Research Reactors Compared with Power Reactors: Nuclear Weapons, Dirty Bombs, and Truck Bombs," *American Behavioral Scientist* 46(6) (2003), pp. 714–726.

30 Zafar (2008).

31 *Ibid.*

32 Anthony Andrews, "Spent Nuclear Fuel Storage Locations and Inventory," *CRS Report for Congress* (Washington, D.C.: Congressional Research Service, December 21, 2004, RS22001).

33 Zafar (2008).

34 *Ibid.*

35 *Ibid.*, p. 19.

36 Quoted in Alexander Shlyakhter, Klaus Stadie, and Richard Wilson, *Constraints Limiting the Expansion of Nuclear Energy* (Washington, D.C.: Global Strategy Council, 1995).

37 Sovacool and Cooper (2008).

38 This text box is taken almost verbatim from Oleg Bukharin, "Security of Fissile Materials in Russia," *Annual Review of Energy and Environment* 21 (1996), pp. 467–496.

39 *Ibid.*

40 Steve Bowman, "Weapons of Mass Destruction: The Terrorist Threat," *CRS Report for Congress* (Washington, D.C.: Congressional Research Service, March 7, 2002, RL31332).

41 Kofi Annan, Address to the General Assembly of the United Nations, March 10, 2005.

42 Government Accountability Office, *Combating Nuclear Smuggling: Corruption, Maintenance, and Coordination Problems Challenge U.S. Efforts to Provide Radiation Detection Equipment to Other Countries* (2006), p. 7 (481 cases); and Rensselaer Lee, "Nuclear Smuggling and International Terrorism: Issues and Options for U.S. Policy," *CRS Report for Congress* (Washington, D.C.: Congressional Research Service, 2002) (426 cases).

43 Kishore Kuchibhotla and Matthew McKinzie, *Nuclear Terrorism and Nuclear Accidents in South Asia* (Washington, D.C.: Stimson Center, February 2004), p. 30.

44 Vijay Sakhuja, "Securing the Nuclear Energy Supply Chain: The Maritime Dimension," Paper presented to the "Emerging Challenges to Energy Security in the Asia Pacific" International Seminar (Chennai, India: Center for Security Analysis, March 16 and 17, 2009).

45   Benjamin K. Sovacool, "Bush Policies Fail to Control Nuclear Smuggling," *Roanoke Times*, January 11, 2004, p. 3.

46   *Ibid.*

47   Zafar (2008), pp. 19–20.

48   *Ibid.*

49   Lovins and Lovins (1982).

50   *Ibid.*

51   Pasternak (2001).

52   *Ibid.*

53   *Ibid.*, p. 145.

54   Pasternak (2001).

55   International Atomic Energy Agency, *Reinforcing the Global Nuclear Order for Peace and Prosperity: The Role of the IAEA to 2020 and Beyond* (Vienna: IAEA, 2008).

56   George Perkovich, "The End of the Nonproliferation Regime?," *Current History* 105(694) (November, 2006), pp. 355–362.

57   Robert Busby, "The United States's Failure to Establish a High-Level Nuclear Waste Storage Facility Is Threatening Its Ability to Effectively Support Nuclear Nonproliferation," *George Washington Journal of International Law and Economics* 30 (1996–1997), pp. 449–480.

58   Federation of American Scientists, *Status of World Nuclear Forces* (Washington, D.C.: FAS, 2010).

59   International Institute for Strategic Studies, *Preventing Nuclear Dangers in Southeast Asia and Australasia* (London: IISS, September 2009), pp. 114–116.

60   Busby (1996–1997).

61   Anthony Salloum, "US–India Nuclear Deal a Non-Proliferation Disaster," *The Toronto Star*, August 21, 2008.

62   John Deutch, "Priority Energy Security Issues," in John Deutch, Anne Lauvergeon, and Widhyawan Prawiraatmadja (eds.), *Energy Security and Climate Change* (Washington, D.C.: Trilateral Commission, 2007), pp. 1–50.

63   Charles J. Hanley, "Uranium-Mining Nations Flout UN on Nuclear Terror," Associated Press, April 4, 2010.

64   Yury Yudin, *Multilateralization of the Nuclear Fuel Cycle: Assessing the Existing Proposals* (Geneva: United Nations Institute for Disarmament Research, 2009), p. xi.

65   Nuclear Energy Agency, *Nuclear Energy in a Sustainable Development Perspective* (Paris: OECD, 2000), p. 9.

66 "Dual Use: Perils of Proliferation," *Jane's Intelligence Digest*, August 12, 2004.

67 Quoted in Pembina Institute, *Nuclear Power and Climate Change* (2007), p. 5.

68 Kuchibhotla and McKinzie (2004).

69 Barnett *et al.* (2006).

70 Source: Kuchibhotla and McKinzie (2004), p. 21.

71 Jim Green, *Nuclear Power: Energy Science Coalition Factsheet* (London: Friends of the Earth, 2003).

72 Richard Weitz, "Global Nuclear Energy Partnership: Progress, Problems, and Prospects," *WMD Insights* (March, 2008).

73 Jungmin Kang, "Analysis of Nuclear Proliferation Resistance," *Progress in Nuclear Energy* 47 (2005), p. 673.

74 Benjamin Sovacool and Saul Halfon, "Reconstructing Iraq: Merging Discourses of Security and Development," *Review of International Studies* 33 (2007), p. 240.

75 *Ibid.*

76 A.E. Farrell, H. Zerriffi, and H. Dowlatabadi, "Energy Infrastructure and Security," *Annual Review of Environment and Resources* 29 (2004), p. 452.

77 Heather Tufts, "The Impact of Uranium Mining on Indigenous Communities," *Peace, Earth, and Justice News* (February 13, 2010).

78 A. Bleise, P.R. Danesi, and W. Burkart, "Properties, Use and Health Effects of Depleted Uranium (DU): A General Overview," *Journal of Environmental Radioactivity* 64 (2003), pp. 93–112.

79 *Ibid.*

80 "How War Debris Could Cause Cancer," *Pharmacy News*, September 6, 2008.

81 *Ibid.*

82 Sakhuja (2009).

83 S. Andseta, M.J. Thompson, J.P. Jarrell, and D.R. Pendergast, "CANDU Reactors and Greenhouse Gas Emissions," in *Proceedings of the 19th Annual Conference, Canadian Nuclear Society, Toronto, Ontario, Canada, October 18–21, 1998* (1998).

84 International Energy Agency, *Environmental and Health Impacts of Electricity Generation: A Comparison of the Environmental Impacts of Hydropower with Those of Other Generation Technologies* (Ontario: IEA Implementing Agreement for Hydropower Technologies and Programs, June 2002).

[85] Matthew Bunn, John P. Holdren, Allison Macfarlane, Susan E. Pickett, Atsuyuki Suzuki, Tatsujiro Suzuki, and Jennifer Weeks, *Interim Storage of Spent Nuclear Fuel: A Safe, Flexible, and Cost-Effective Near-Term Approach to Spent Fuel Management* (Cambridge, MA and Tokyo: A Joint Report from the Harvard University Project on Managing the Atom and the University of Tokyo Project on Sociotechnics of Nuclear Energy, June 2001).

[86] Allison Macfarlane, "Interim Storage of Spent Fuel in the United States," *Annual Review of Energy and Environment* 26 (2001), pp. 201–235.

[87] Fred Dilger and Robert Halstead, "The Next Species of Trouble: Spent Nuclear Fuel Transportation in the United States, 2010–2048," *American Behavioral Scientist* 46 (2003), pp. 796–811.

[88] *Ibid.*

[89] *Ibid.*

[90] Robert Nadelson, "After MOX: The Contemporary Shipment of Radioactive Substances in the Law of the Sea," *International Journal of Marine and Coastal Law* 15(2) (2000), pp. 193–244.

[91] David B. Dixon, "Transnational Shipments of Nuclear Materials by Sea: Do Current Safeguards Provide Coastal States a Right to Deny Innocent Passage?," *Journal of Transnational Law & Policy* 16 (2006/2007), pp. 75–76.

[92] John M. Van Dyke, "The Legal Regime Governing Sea Transport of Ultra-Hazardous Radioactive Materials," *Ocean Development & International Law* 33 (2002), pp. 78–79.

[93] The following paragraphs are taken mostly from Bernard D. Cole, *Sea Lanes and Pipelines: Energy Security in Asia* (London: Praegar, 2008).

[94] *Ibid.*, p. 87.

[95] International Institute for Strategic Studies (2009), p. 43.

[96] Dixon (2006/2007), pp. 73–99; and Sakhuja (2009).

[97] Dixon (2006/2007).

[98] Nadelson (2000).

[99] Dixon (2006/2007), p. 79.

[100] Dixon (2006/2007).

[101] Sakhuja (2009).

[102] Van Dyke (2002), pp. 77–98.

[103] A. Blowers and P. Leroy, "Power, Politics and Environmental Inequality: A Theoretical and Empirical Analysis of the Process of 'Peripheralisation,'" *Environmental Politics* 3(2) (Summer, 1994), pp. 197–228.

[104] C. Michael Rasmussen, "Getting Access to Billions of Dollars and Having a Nuclear Waste Backyard," *Journal of Land Resources and Environmental Law* 18 (1998), pp. 335–367.

[105] James M. Jasper, "Gods, Titans, and Mortals: Patterns of State Involvement in Nuclear Development," *Energy Policy* (July, 1992), pp. 653–659.

[106] Benjamin K. Sovacool and Scott Valentine, "The Socio-Political Economy of Nuclear Energy in China and India," *Energy* 35 (September, 2010), pp. 3803–3813.

[107] Sulfikar Amir, "Challenging Nuclear: Antinuclear Movements in Postauthoritarian Indonesia," *East Asian Science, Technology, and Society* (September 11, 2009), pp. 1–22.

[108] Dominique Finon and Carine Staropoli, "Institutional and Technological Co-Evolution in the French Electronuclear Industry," *Industry and Innovation* 8(2) (2001), pp. 179–199.

[109] Phillip A. Greenberg, "Safety, Accidents, and Public Acceptance," in John Byrne and Steven M. Hoffman (eds.), *Governing the Atom: The Politics of Risk* (London: Transaction Publishers, 1996), pp. 127–175.

[110] *Ibid.*

[111] David J. Rose, "Energy and History," *American Heritage* 32 (June/July, 1981), pp. 79–80.

[112] Abby Luby, "Nuclear Steam Leak Intentional: Response to Indian Point Plant Shutdown," *Daily News*, January 8, 2010, p. 1.

[113] See Frans Berkhout, *Radioactive Waste: Politics and Technology* (London: Routledge, 1991); A. Blowers, D. Lowry, and B. Solomon, *The International Politics of Nuclear Waste* (New York: Macmillan, 1991); A. Blowers, "Nuclear Waste and Landscapes of Risk," *Landscapes Research* 24(3) (1999), pp. 241–264; and Peter Stoett, "Toward Renewed Legitimacy? Nuclear Power, Global Warming, and Security," *Global Environmental Politics* 3(1) (2003), pp. 99–116.

[114] Darrin Durant, "Buying Globally, Acting Locally: Control and Co-Option in Nuclear Waste Management," *Science and Public Policy* 34(7) (August, 2007), pp. 515–528.

7

Energy Efficiency and Renewable Energy: "The Fire Extinguisher"

Marion King Hubbert, the geophysicist who predicted that American oil production would peak in about 1970, often remarked that it would be incredibly difficult for people living now, having been accustomed to exponential growth in energy consumption, to assess the transitory nature of conventional sources of energy.[1] Hubbert argued that proper reflection could only happen if one looked at a time scale of 10,000 years. On such a scale, Hubbert thought that the complete cycle of the world's exploitation of fossil and nuclear fuels would encompass perhaps 1,100 years, with the principal segment of this cycle covering 300 years. Indeed, Chapter 3 showed that, at current rates of consumption, the world has less than 83 years of uranium supply left. From now until the point where they truly become depleted, coal, natural gas, oil, and uranium will always be subject to perceived scarcity, market speculation, manipulation, and price gouging.

Energy efficiency and renewable flows of energy, by contrast, have no foreseeable end. We need not, in other words, commit ourselves to a nuclear power system characterized by intractable technical, social, economic, and political challenges. As German parliamentarian Hermann Scheer put it, "our dependence on fossil fuels amounts to global pyromania . . . and the only fire extinguisher we have at our disposal is renewable energy."[2] This chapter outlines how energy efficiency practices and technologies as well as renewable energy systems can respond to the energy-related challenges facing society better than nuclear reactors. Energy efficiency and demand-side management — doing more with less, lowering levels of energy consumption by substituting fuels and technologies and altering consumer behavior — is often the cheapest and quickest way to address increases in demand for energy services. Renewable power generators — unlike nuclear power plants which rely on uranium mining or reprocessing — utilize sunlight, wind, falling water, biomass, waste, or geothermal heat to produce electricity from fuels that are mostly free for the taking. These "fuels" also happen to be in great abundance in every country in the world, and thus offer a way to make electricity sectors less susceptible to supply chain interruptions and shortages. Vikram Budhraja, former President of Edison Technology Solutions, noted that "the beauty of renewable energy" is that "it has no fuel input," and that its fuel price remains relatively constant no matter how much the market price of oil, gas, coal, or uranium changes.[3]

Energy Efficiency

Before readers start to despair, put this book down and vow never to read anything this author writes ever again — there is hope. There is one resource widely abundant around the world, in every country, and according to some studies one that offers more potential than any other known source of energy. Unlike the next generation of nuclear reactors, it is already commercially available and is ready to be utilized without the need for subsidies or further research. It is not capital-intensive, does not consume water, emits very low amounts of greenhouse gases, does not melt down, and does not rely on a depletable fuel. What is this wondrous resource? Energy efficiency and demand-side management.

Energy efficiency can include practices as diverse as switching from conventional coal power plants to combined heat and power units, lowering thermostats, better maintaining industrial boilers, and walking or cycling to work instead of driving. It involves light bulbs that require less power, weather stripping around doors and windows, more efficient industrial motors and pumps, better HVAC (heating, ventilation, and air-conditioning) systems, and improved vehicle mileage. Studies have found, for example, that every dollar invested in energy efficiency:

- Mitigates against uncertainty and reduces load, wear, and maintenance needs on the entire fossil fuel chain, even in times when reliability problems are not anticipated by system managers;
- Depresses the costs of locally used fuels such as uranium, oil, coal, and natural gas;
- Reduces demand across peak hours, the most expensive times to produce power;
- Lessens costly pollutants and emissions from generators;
- Improves the reliability of existing generators;
- Moderates transmission congestion problems; and
- Operates automatically through customers coincident with the use of underlying equipment or load, meaning efficiency measures are always "on" without any delay or needed intervention by system operators to schedule or purchase the resource.[4]

The potential of these sorts of actions is so great that aggressive investments in energy efficiency in industrialized countries such as the United States, which has already implemented national energy efficiency programs, would eliminate the need to construct more than 1,300 power plants in the next 20 years with net savings.[5] They also pay for themselves quite quickly. One assessment of 41 energy efficiency projects completed in the industrial sector of the US found that they recovered the cost of implementation in slightly more than two years and then yielded an aggregate US$7.4 million in savings every year thereafter.[6] On a global scale, the International Energy Agency reviewed large-scale energy efficiency programs and found that they saved electricity at an average cost of 3.2 cents/kWh, well below the cost of supplying electricity independent of its source.[7]

In the US electricity sector, studies have shown that cost-effective energy efficiency measures could reduce national consumption by an astounding 30–75%. These measures are cheaper to implement than purchasing any form of electricity supply, and could save up to three-quarters of the country's power bill.[8] Another assessment by the Federal Energy Regulatory Commission looking at customer demand response to hourly market-based retail prices in the UK and the US found that it could reduce electricity loads by thousands of megawatts, lower summer peak prices by 19%, and produce energy cost savings from US$300 million to US$1.2 billion per year.[9] Put directly in the context of nuclear reactors, another study noted that cost-effective energy efficiency resources could save the need to build more than 100 new nuclear power plants between 2010 and 2030.[10] A comprehensive report from McKinsey looked at the maximum potential of an assortment of technological options to abate greenhouse gases (presuming a tax of €60 per ton of CO_2 was in place), and found that a host of residential and industrial energy efficiency options — including geothermal energy, bioelectricity, small-scale hydro-electricity, and building efficiency — were far more cost-effective than nuclear power.[11] Its findings are depicted in Figure 1.

The attractiveness of energy efficiency — namely, its ability to represent "low-hanging fruit" time and time again — probably explains

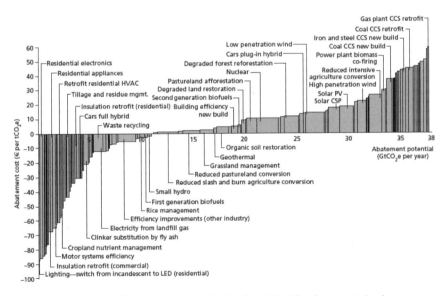

Figure 1:   McKinsey Cost Curve for Carbon Dioxide Abatement Options

why energy efficiency efforts have saved more energy, thus far, many times over what nuclear reactors produce each year. Looking at energy as a whole (and not just electricity), total primary energy use per capita in the US in 2000 was almost identical to energy use per capita in 1973. Over the same 27-year period, economic output (measured in terms of GDP per capita) increased by 74% while national energy intensity (or energy use per unit of GDP) fell by 42%. About three-fifths of this decline can be attributed to energy efficiency improvements. If the US had not dramatically reduced its energy intensity over the years, consumers would have spent at least US$530 billion more on energy purchases in 2007 (or more than US$1.4 billion *every day*).[12] In other words, as Figure 2 shows, when applied to the electricity industry, the gains made by energy efficiency outdo *every single source of electricity generation* today, including coal, natural gas, and nuclear. Energy intensity has continued to fall in the US, declining by 8% from 2001 to 2005 while GDP grew by 3%.[13] This trend holds true not just for the US, but for most industrialized countries. Figure 3 reveals findings from a

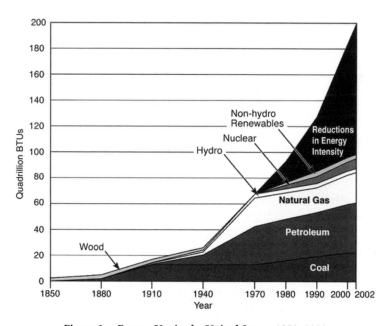

Figure 2:   Energy Use in the United States, 1850–2002

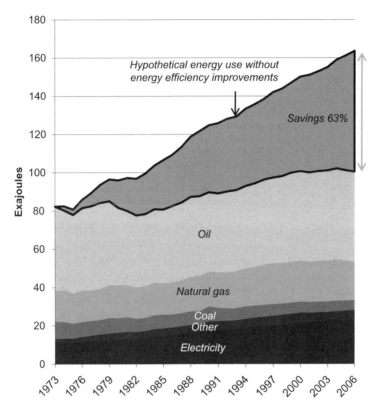

Figure 3: Energy Efficiency Savings in 11 OECD Countries, 1973–2006

comprehensive assessment of energy use in 11 Organisation for Economic Co-operation and Development (OECD) countries, which noted that energy efficiency practices saved more energy (in exajoules) than any other source of electricity supply.

What has accounted for the dramatic improvements in efficiency and reductions in energy intensity? Equipment standards, required by many governments in the 1970s after the oil shocks, required manufacturers to test and standardize the energy usage of appliances. These standards forced manufacturers to improve the energy efficiency of gas furnaces by 25% between 1972 and 2001; central air conditioners, by 40%; and refrigerators, by 75%. From 2000 to 2015, adherence to appliance standards will save consumers around US$85 billion on a net present value basis in the US.[14] Individuals have done their part, too, by purchasing more fuel-efficient appliances, insulating and weatherproofing their homes, and adjusting thermostats to reduce energy consumption. Businesses have retrofitted their buildings with more efficient heating and cooling equipment and also installed energy management and control systems, accounting for a decline of 25% of energy use per square foot of commercial building space. Factories have adopted more efficient manufacturing processes and employed more efficient motors for conveyors, pumps, fans, and compressors. Some electric utilities discovered that they could save electricity cheaper than the cost of operating existing plants, meaning efficiency can improve cash flow by displacing operating costs, thus appeasing investors and saving consumers money at the same time.[15] Southern California Edison ran a huge energy efficiency campaign that made money even when credited solely with the value of unburned fuel.[16] The UK passed an "Energy Efficiency Commitment" scheme that obligated all electricity and gas suppliers to save electricity, and Italy passed two ministerial decrees related to energy efficiency that ended up saving electricity at a cost of 1.7 cents/kWh.[17]

Renewable Energy

Although industrialized countries can best meet most of their energy challenges with energy efficiency, developing countries still have to build power plants from the ground up. Nonetheless, both sets of countries will inevitably reach a point where new sources of electricity supply may be needed. When this occurs, after energy efficiency options have been exhausted, the best systems to build and install are all renewable power systems.

Renewable Energy Basics

Operators generally divide renewable power systems into six types, as depicted in Table 1: wind turbines (onshore and offshore), solar energy (including solar photovoltaic panels, solar thermal systems, and concentrated solar power), geothermal energy (conventional and advanced),

Table 1:   **Renewable Power Generators and Associated Fuel Cycles**

Source	Description	Fuel	Size of Individual Units
Onshore wind	Wind turbines capture the kinetic energy of air and convert it into electricity via a turbine and generator	Wind	1.5 kW to 2.5 MW
Offshore wind	Offshore wind turbines operate in the same manner as onshore systems, but are moored or stabilized to the ocean floor	Wind	750 kW to 5 MW
Solar PV	Solar photovoltaic cells convert sunlight into electrical energy through the use of semiconductor wafers	Sunlight	1 W to 100 MW
Solar thermal	Solar thermal systems use mirrors and other reflective surfaces to concentrate solar radiation, utilizing the resulting high temperatures to produce steam that directly powers a turbine; the three most common generation technologies are parabolic troughs, power towers, and dish-engine systems	Sunlight	5 kW to 320 MW
Geothermal (conventional)	An electrical-grade geothermal system is one that can generate electricity by means of driving a turbine with geothermal fluids heated by the earth's crust	Hydrothermal fluids heated by the earth's crust	25–1,400 MW
Geothermal (advanced)	Deep geothermal generators utilize engineered reservoirs that have been created to extract heat from water while it comes into contact with hot rock and then returns to the surface through production wells	Hydrothermal fluids heated by the earth's crust	10–1,500 MW
Biomass (combustion)	Biomass generators combust biological material to produce electricity, sometimes gasifying it prior to combustion to increase efficiency	Agricultural residues, wood chips, forest waste, energy crops	20–50 MW
Biomass (landfill gas)	These biomass plants generate electricity from landfill gas and anaerobic digestion	Municipal and industrial waste, trash	30 kW to 10.5 MW
Hydroelectric	Hydroelectric dams impede the flow of water and regulate its flow to generate electricity	Water	200 kW to 6,809 MW
Ocean power	Ocean, tidal, wave, and thermal power systems utilize the movement of ocean currents and the heat of ocean waters to produce electricity	Saline water	N/A

biomass energy (including agricultural waste, energy crops, trash, and landfill gas), hydroelectricity (big and small), and ocean power.

Wind turbines convert the flow of air into electricity. They are most competitive in areas with stronger and more constant winds, such as off-shore locations or regions of high altitude.

Solar photovoltaic (PV) cells, also called "flat-plate collectors," convert sunlight into electrical energy through the use of semiconductor wafers, and are often used in arrays and integrated into buildings. Solar thermal systems, also called "concentrated" or "concentrating" solar power, use mirrors and other reflective surfaces to concentrate solar radiation, utilizing the resulting high temperatures to produce steam to then power a turbine.

An electrical-grade geothermal system is one that can generate electricity by means of driving a turbine with geothermal fluids heated by the earth's crust.

Biomass generators combust agricultural residues, wood chips, forest waste, energy crops, municipal and industrial waste, and trash to produce electricity. Biomass generation also includes advanced combustion techniques such as biomass gasification (in which the biomaterial is gasified to increase efficiency prior to its combustion), co-firing (in which biomass burns with another fuel, such as coal or natural gas, to increase its density), as well as electrical generation from landfill gas and anaerobic digestion.

Two types of hydroelectric facilities exist: large-scale facilities that consist of a dam or reservoir impeding water and regulating its flow, and run-of-river plants that create a small impoundment to store a day's supply of water. Smaller hydroelectric systems, also referred to as "run-of-the-mill," "micro-hydro," or "run-of-the-river" hydropower, consist of a water conveyance channel or pressured pipeline to deliver water to a turbine or waterwheel that powers a generator, which in turn transforms the energy of flowing water into electricity; the diverted water is then sent almost immediately back into the flow of the original source. Because they operate on a much smaller scale, use smaller turbines, and require much less water, run-of-the-mill hydro plants escape many of the challenges raised by their larger counterparts.

The category of electricity known as ocean power includes shoreline, near-shore, and offshore wave extraction technologies and ocean thermal energy conversion (OTEC) systems. Because they are a much newer technology than other renewables, comprehensive cost analyses and product reviews are limited. Since ocean power plants do not currently exist in the commercial sector, I do not discuss them in this chapter.

Advances in design, operation, and maintenance now enable these types of power technologies to generate electricity more reliably than nuclear plants. Geothermal, bioelectric, and hydroelectric plants have long provided reliable baseload power in the same fashion as nuclear plants. One 2008 assessment of hydroelectric power in the US found that, by looking at just four possible resources (i.e. constructing new but smaller-scale dams, upgrading existing facilities, adding power generators to non-hydroelectric dams, and commercializing hydrokinetics), 58,882–311,202 MW of installed baseload capacity was available. That amount is equivalent to between 50 and 300 new nuclear reactors, and it already takes into account restrictive environmental standards.[18]

Previously intermittent sources such as wind and solar can also displace nuclear resources. No less than nine recent studies have concluded that the variability or intermittency of wind and solar resources becomes easier to manage the more they are deployed and interconnected, and not the other way around, as some utilities suggest.[19] This is because wind and solar plants help grid operators handle major outages and contingencies elsewhere in the system, since they generate power in smaller increments that are less damaging than unexpected outages from large plants. Most studies critiquing wind and solar energy look only at individual generators, but not at the system-wide effects of integrated wind and solar farms. Correlations between power swings drop substantially as more wind farms are integrated — a process known as geographical smoothing — and a wider geographic area also enables a larger pool of energy efficiency efforts to abate intermittency.[20] The author and a colleague from the Georgia Institute of Technology found that, when coupled with a rigorous energy efficiency and demand management program, solar panels could completely displace the electricity currently coming from the 2-GW Indian Point nuclear facility in New York.[21]

Apart from connecting intermittent renewables of the same type, different renewables can be integrated together (or with energy efficiency measures and technologies) to create very reliable hybrid systems.

Installing wind turbines at geothermal power plants creates effective base-load systems, as wind data already exist at plant locations to site cooling towers and plant designs allow for suitable spare land. These plants can rely on geothermal electricity to back up or offset any unexpected short-falls in wind.[22] Similarly, wind farms can be coupled with biomass plants to completely eliminate their intermittency, using agricultural wastes and residues, methane from landfills, energy crops, and trash as sources of fuel.[23]

A far more extensive hybrid system, called the "Combined Power Plant" or the "Renewable-Energy Combined Cycle Power Station" (*kombikraftwerk*), exists in Germany. Operated by Schmack Biogas AG, SolarWorld AG, and Enercon, this Combined Power Plant relies on an integrated network of 36 wind, solar, biomass, and hydropower installa-tions spread across Germany. Wind and solar units generate electricity when those resources are available, while a collection of biomass and bio-gas plants and a pumped hydro facility make up the difference when they are not. The system can immediately adapt to a shortfall in any one resource by drawing on the others. As of early 2009, the 23.2-MW Combined Power Plant consisted of 11 wind turbines at three separate wind farms, 23 distributed solar systems, 4 combined heat and power bio-gas units, and a pumped hydro storage plant, all linked via central control. In 2008, the facility produced 41.1 GWh of electricity without a single interruption of supply — enough electricity for 12,000 households in Schwäbisch Hall. This project shows, quite clearly, that a combination of different renewable energy technologies can potentially cover the entire electricity demand in Germany. The project size was chosen to represent German electricity demand on a scale of 1 to 10,000. The Combined Power Plant also lowered the region's dependence on oil and natural gas, and produced no direct greenhouse gas emissions.[24]

A similar hybrid system exists in the Saxony-Anhalt state of Germany near the Harz district. There, 6 MW of wind are connected to an 80-MW pumped hydro facility, which is used to back up wind output by pumping water up when wind is available and using gravity to power two 40-MW turbines to balance the system when wind is not available. This wind–hydro system is in the process of being integrated with distributed solar power plants in the village of Dardesheim, six biogas systems, and a large

5-MW cogeneration unit fueled by recycled vegetable oil. The resulting wind–hydro–solar–biogas–vegetable oil facility, integrated via a digital control station, is expected to provide about 500 million kWh of electricity to a region that consumes only 800 million kWh, meaning it will meet two-thirds of all electricity demand.[25]

These integrated and reliable renewable energy systems are not limited to Germany or Europe. In Zambia, an interconnected solar–biomass–micro-hydro network will generate baseload electricity for a collection of local villages. The combined system will include a collection of distributed solar panels, one biomass power plant, and one micro-hydroelectric station, with a collective output of 2.4 MW; and it was expected to begin operation in 2010.[26] In Cuba, a hybrid biomass gasification power plant, four distributed biogas plants, and one wind farm will have a rated capacity of 11 MW and begin generating baseload electricity for Isla de la Juventud in 2011.[27] In the village of Xcalak, Mexico, 234 solar panels have been integrated with 36 batteries, 6 wind turbines, a 40-kW inverter to convert direct-current (DC) power to alternating-current (AC) power, and a sophisticated control system. The system has so far displaced the need to construct a US$3.2 million transmission line extension, and in its first year of operation proved more reliable than the diesel generators that it replaced (although one is still installed as a backup, just in case).[28]

At the commercial headquarters of the Rocky Mountain Institute, solar thermal and solar PV panels have been integrated with energy efficiency techniques so that renewable energy meets 99% of the building's demand for energy services. Using a passive design, super-efficient windows, state-of-the-art insulation and ventilation, a solar water heater, two exterior solar PV tracking systems, and a collection of stationary flat-plate solar panels integrated into the roof, the building needs no conventional heating system. The building is so efficient that, when one of the owners wants to turn up the heat, he remarked that it can be harnessed "from a 50-watt dog, adjustable to 100-watt by throwing a ball."[29]

Similarly, a California prison has integrated about 3 acres of solar panels with net metering, energy efficiency, and storage systems to not only meet the facility's energy needs but also export electricity back to the grid.

The entire system is designed to collect, store, and save electricity in the morning and at night; and to then sell it back to the grid during times of peak demand, when it is worth the most. The system paid for itself within the first year of operation, and has since brought hundreds of thousands of dollars in additional savings.[30]

Looking to the future, as the performance of batteries improves and their costs drop, they, too, could begin to back up large amounts of solar and wind power. Likewise, the greater use of plug-in hybrid electric vehicles and vehicle-to-grid technologies can also enhance the competitiveness of intermittent renewables, as these innovations would enable automobiles to store energy from wind and solar sources that could then be recalled when needed.[31] The potential resource base for tapping batteries in vehicles to store energy from renewables is staggering. Placing just a 15-kW battery in each of the existing 191 million automobiles in the US would create 2,865 GW of equivalent electricity capacity, if all of the vehicles supplied power simultaneously to the grid — a capacity more than twice that of the entire existing electricity industry (which has slightly more than 1,000 GW of installed capacity).[32]

Lastly, when interconnecting wind and solar farms or creating hybrid systems is not practicable or possible, intermittent renewable systems can be integrated with energy storage technologies to eliminate their intermittency. Although batteries are the most common type of commercially available storage technology on the market, they are also the most expensive and tend to store only small amounts of energy. Three other types of storage systems — pumped hydro, compressed air, and molten salt — offer a cheaper alternative and can handle larger amounts of capacity. Each of these three options tends to add only 0.6–5.1 extra cents per kWh to the levelized cost of renewable electricity generation.

As we shall see, these types of renewable energy systems — large or small; grid-connected or off-grid; and by themselves, spatially integrated, configured as hybrid systems, or coupled with storage technologies — have immense advantages compared to nuclear reactors. They (1) are less expensive, (2) use domestically available fuels, (3) preserve the land and produce little waste, (4) need less water, (5) emit fewer greenhouse gases, and (6) improve safety and security. The remainder of this chapter covers each dimension in detail.

Cost

In contrast to gargantuan nuclear units, most renewable power technologies have quicker construction lead times, taking just a few months or years to implement. There is no need for mining, milling, and leaching uranium, enriching and reprocessing fuel assemblies, or permanently storing radioactive waste. The quicker lead times for renewables enable a more accurate response to load growth, and minimize the financial risk associated with borrowing hundreds of millions of dollars to finance plants for ten or more years before they start producing a single kilowatt-hour of electricity. Florida Power & Light (FPL) claimed that it can take a new wind farm from groundbreaking to commercial operation in as little as 3–6 months. In 2005, Puget Sound Energy (PSE) proved that FPL's boast was achievable in practice, when it brought 83 1.8-MW wind turbines at its Hopkins Ridge Wind Project from groundbreaking to commercial operation in exactly six months and nine days.[33] Moreover, wind turbines are not the only technology that can achieve quick lead times; in Nevada, Ormat Nevada Inc. commissioned a 20-MW geothermal power plant only eight months after groundbreaking.[34]

Solar panels can be built in various sizes, placed in arrays ranging from watts to megawatts, and used in a wide variety of applications (including centralized plants, distributed substation plants, grid-connected systems for home and business use, and off-grid systems for remote power use). PV systems have long been used to power remote data relay stations, which are critical to the operation of supervisory control and data acquisition systems used by electric and gas utilities and government agencies. Solar installations may require even less construction time than wind or geothermal facilities, since the materials are prefabricated and modular. The Partnership for Advancing Technology in Housing recently conducted a case study of a typical PV system-powered home, finding that it required only a two-month lead time for the panels.[35]

Utilities and investors can cancel modular plants more easily, so abandoning a project is not a complete loss, and the portability of most renewable systems means that recoverable value exists should the technologies need to be resold as commodities in a secondary market. A half-completed nuclear plant is still a construction project, perhaps even a liability; whereas a half-completed wind plant can produce electricity from its already-installed wind turbines. Smaller units with shorter lead times reduce the risk of buying a technology that becomes obsolete even before it is installed. Also, quick installations can better exploit rapid learning, as "many generations of product development can be compressed into the time it would take simply to build a single giant power plant."[36] In addition, outage durations tend to be shorter than those from larger plants; and repairs take less money, time, and skill. As one study concluded, "technologies that deploy like cell phones and personal computers are faster than those that build like cathedrals. Options that can be mass-produced and adopted by millions of customers will save more carbon and money sooner than those that need specialized institutions, arcane skills, and suppression of dissent."[37] Smaller projects, too, involve less money and interest. A quarter or more of the total construction cost for large power plants tends to be the interest paid on construction capital before commissioning, but economies of scale are nonexistent in interest rates: bankers often charge the same rate for large and small loans.[38]

The United Nations recently calculated, in a study utilizing 2007 data collected from dozens of countries, that renewable power sources can produce incredibly cheap power without subsidies.[39] At the low end of the range, Table 2 shows that hydroelectric, geothermal, wind, and biomass

Table 2: Levelized Cost of Electricity (LCOE) for New Renewable and Nuclear Power Plants

Source of Electricity	Nominal LCOE (2007 USD, cents/kWh)
Hydroelectric	3–7
Geothermal	4–7
Wind	5–12
Bioelectric	5–12
Solar thermal	12–18
Nuclear	18–30
Solar PV	20–80

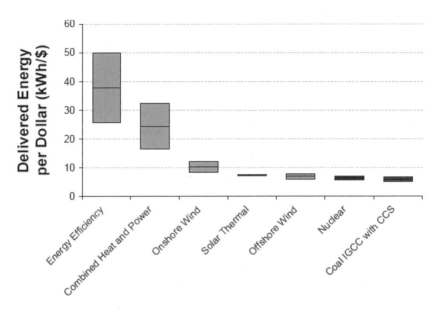

Figure 4: Delivered Cost of Energy per Dollar of Investment for Selected Low-Carbon Energy Technologies

energy can all generate electricity for 5 cents/kWh or less.[40] Such numbers were again confirmed with a 2009 estimate which found that energy efficiency, combined heat and power, onshore and offshore wind power, and even concentrated solar power all deliver total energy at costs much cheaper than new nuclear units.[41] These numbers are presented in Figure 4. Without additional subsidies, most renewable power sources, with their "intermittent" or "low" capacity factors, are already cost-competitive with conventional systems. Their progress is all the more impressive, considering that these technologies reached such a point while receiving only a small fraction of the subsidies set aside for conventional systems.

Renewable fuels have at least two economic advantages over nuclear power plants (and fossil-fueled power stations). First, they are often indigenous and free. For the most part, every kilowatt-hour generated from sunlight or wind does not compete with the sunlight or wind available elsewhere. Countries need not expend considerable resources to secure renewable supplies. Put another way, a ton of coal or a barrel of oil used by one community cannot be used by another; whereas renewable resources, because they are non-depletable, do not force such geopolitical tradeoffs. Moreover, the fuel cost for renewables can be known for 20 years into the future — something that cannot be said about conventional technologies, where predicting spot prices in the future is about as easy as reading crystal balls or practicing astrology.

Second, renewable energy resources keep money in the local economy. Studies have shown that, for every dollar spent by consumers on fossil fuels, some US$0.85 leaves the local economy.[42] About 93% of the fuel distributed to electric utilities in the Southeastern part of the United States in 2005 had to be imported from other states and countries. The region lost US$7.3 billion of revenue paying for coal (with more than US$700 million going to coal from Colombia, Venezuela, and Poland) and US$1.1 billion importing uranium over the course of one year. A study conducted in California found that every dollar invested in solar thermal plants resulted in US$1.40 of gross state output, while dollars invested in fossil fuels merely transferred revenue out of the region. The state of Arizona estimated that 79.4 cents of every dollar spent on fossil fuel and 56.5 cents spent on electricity are completely diverted from the local economy.[43] It is a perverse system indeed that currently spends billions of dollars each year to pay for foreign uranium and then suggests that the answer to high prices is to explore for more uranium, making countries even more dependent. By contrast, if a country such as the US received 20% of its electricity from renewable resources, rural landowners would receive up to US$562 million per year in payments from wind leases — money that would stay in the economy rather than going abroad.[44]

Fuel Availability

The earth receives radiation from the sun in a quantity far exceeding humanity's needs. By heating the planet, the sun generates wind and creates waves. The sun powers the evapotranspiration cycle, allowing for the generation of power from hydroelectric sources. Plants photosynthesize, creating a wide range of biomass products. A comprehensive study undertaken by the US Department of Energy (DOE) calculated that 93.3% of all domestically available energy in the US is in the form of wind, geothermal, solar, and biomass energy. The country is literally the Saudi Arabia of

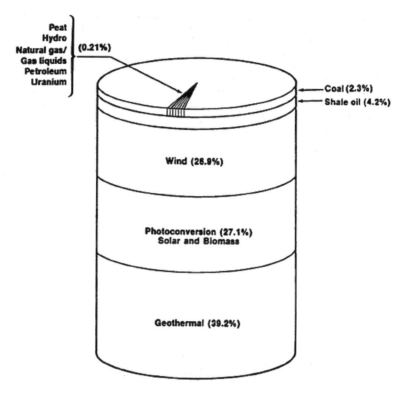

Figure 5:   Domestic Energy Resources and Reserves in the United States

renewable resources. In fact, the DOE estimated in Figure 5 that the total amount of renewable resources found within the country amounts to a total resource base equivalent to 657,000 billion barrels of oil — more than 46,800 times the national rate of energy consumption per year.[45] Amazingly, this estimate was validated by researchers at the US Geological Survey, Oak Ridge National Laboratory, Pacific Northwest National Laboratory, Sandia National Laboratories, National Renewable Energy Laboratory, Colorado School of Mines, and Pennsylvania State University.

According to published, nonpartisan, and peer-reviewed estimates from a collection of scientific studies presented in Table 3, and not estimates from manufacturers or trade associations, assuming the utilization of existing, commercially available technologies, the US had 3,730,721 MW of achievable renewable energy potential in 2010. Within this

Table 3:   Renewable Energy Potential (by source) for the United States[47]

	Electricity Generation (thousand kWh) in 2006	Grid-Connected Installed Capacity (MW) in 2007	Potential (MW)
Onshore wind	25,781,754	12,600	1,497,000
Offshore wind	0	0	791,000
Solar PV	505,415	624	710,000
Solar thermal/CSP	N/A	354	98,000
Geothermal	14,842,067	3,100	2,800
Biomass (combustion)	50,064,892	9,733	465,000
Biomass (landfill gas)	5,509,189	539	1,370
Hydroelectric	288,306,061	80,000	165,551
Total	385,009,378	106,950	3,730,721

estimate, two numbers become significant: first, renewable resources have the capability to provide 3.7 times the total amount of installed electricity capacity operating in 2008; and second, the country has harnessed only 2.9% of the potential energy to be found in the nation's available renewable resources. As the National Research Council recently concluded in 2009:

> [F]or the time period from the present to 2020, there are no current technological constraints for wind, solar photovoltaics and concentrating solar power, conventional geothermal, and biopower technologies to accelerate deployment. . . . Renewable resources available in the United States, taken collectively, can supply significantly greater amounts of electricity than the total current or projected domestic demand.[46]

The same is true for Asia. Just five regions in Asia — the member countries of the Association of Southeast Asian Nations (ASEAN), China, India, Japan, and South Korea — have an achievable renewable power potential of 2,646.5 GW, which is more than 2.5 times the amount of power expected to be utilized by these areas in 2010 (see Table 4). While "achievable" potential does not necessarily mean economic potential (it refers to what could be built today, regardless of its cost), Southeast Asia

Table 4: Potential for Commercially Available Renewable Electricity Generators in ASEAN, China, India, Japan, and South Korea

Country/Region	Projected Capacity Needed by 2010 (GW)	Installed Renewable Power Capacity (GW) in 2006	Achievable Renewable Power Capacity (GW) by 2010	
ASEAN	90	14	520.8	Wind (138.7) Solar (11) Geothermal (30) Hydroelectric (254.2) Biomass (86.9)
China	500	75	1,475	Onshore wind (253) Offshore wind (750) Solar PV (0.35) Solar thermal (60) Geothermal (5.8) Hydroelectric (400) Biomass (5.5)
India	140	3.7	245.6	Wind (47) Solar (50) Geothermal (10.6) Hydroelectric (15) Biomass and bagasse (73) Ocean (50)
Japan	255	25	324.4	Wind (222) Solar (4.8) Geothermal (70) Hydroelectric (26.5) Agricultural residue (1.1)
South Korea	56	3	80.5	Wind (53) Solar (4.3) Geothermal (14) Hydroelectric (6.9) Biomass (2.3)
Total	1,041	121	2,646.5	

does boast significant hydroelectric and geothermal reserves; China, India, and Japan possess immense reserves of biomass and wind power; and South Korea has substantial wind and solar energy resources. Perhaps more astonishingly, Table 4 shows that regulators have so far installed only 4.7% of this achievable potential.

The US and Asia are not representative of all countries; nevertheless, global numbers for the resource potential of renewables tell the same story. Solar, wind, and biomass resources exist in all countries; hydroelectric resources, in most countries; and geothermal resources, in many countries around the Pacific Rim. Excluding biomass energy, and looking at just solar, wind, geothermal, and hydroelectric energy, the world has roughly 3,429,685 TWh of potential — about 201 times the amount of electricity the world consumed in 2007 (see Table 5). So far, only 3,082.4 TWh of that capacity (less than 0.09%) has been tapped.

Land Use and Waste

Renewable power sources require less land area than nuclear power plants and facilities, and most of the land they occupy is still "dual-use." When configured in large centralized plants and farms, wind and solar technologies use around 10–78 km^2 of land per installed GW per year; whereas traditional plants can use more than 100 km^2 of land per year to produce the same amount of electricity. In open and flat terrain, newer large-scale wind plants require about 60 acres per MW of installed capacity, but the

Table 5:   Renewable Energy Potential (by source) for the World[48]

Technology	Available Energy (TWh/year)	Electrical Potential (TWh/year)	Current Electricity Generation (TWh/year)
Solar PV	14,900,000	3,000,000	11.4
Concentrated solar power	10,525,000	4,425	0.4
Wind	630,000	410,000	173
Geothermal	1,390,000	890	57.6
Hydroelectric	16,500	14,370	2,840
Total	**27,461,500**	**3,429,685**	**3,082.4**

amount drops to as little as 2 acres per MW for hilly terrain. While 60 acres may sound like a lot, only 5% (or three acres) of this area is actually occupied by turbines, access roads, and other equipment; 95% remains free for other compatible uses such as farming or ranching. Only a small fraction of contiguous land in the country, ranging from 0.11% to 0.26%, would be needed to supply 20% of the nation's electricity from wind energy; and of that land, more than 98% would be available for other uses.[49] At the High Winds Project in Solano, California, for example, eight different landowners host 90 separate 1.8-MW wind turbines that total 162 MW of electricity capacity, but are still able to use about 96% of farmland around and between the turbines.

When integrated into building structures and facades, solar PV systems would require no new land at all. The California Exposition Center in Sacramento, California, for example, has fully integrated a 540-kW PV system into a parking lot. Indeed, the National Renewable Energy Laboratory concluded that "a world relying on PV would offer a landscape almost indistinguishable from the landscape we know today."[50] The Energy Policy Initiatives Center at the University of San Diego recently estimated that the city could construct a 1,532-GWh solar PV system by relying only on available roof areas downtown. Moreover, the Worldwatch Institute, an independent think tank, noted that "solar power plants that concentrate sunlight in desert areas require 2,540 acres per billion kWh. On a lifecycle basis, this is less land than a comparable coal or hydropower plant [generating the same amount of electricity] requires."[51]

High-yield food crops leach nutrients from the soil, but the cultivation of biomass energy crops on degraded land can help stabilize soil quality, improve fertility, reduce erosion, and improve ecosystem health. Perennial energy crops improve land cover and enable plants to form an extensive root system, adding to the organic matter content of the soil. Agricultural researchers in Iowa, for instance, discovered that planting grasses or poplar trees in buffers along waterways helped to capture runoff from corn fields, making streams cleaner. Prairie grasses, with their deep roots, build up topsoil, putting nitrogen and other nutrients into the ground. Twigs and leaves decompose in the field after harvesting, enhancing soil nutrient composition. Biomass crops can also create better wildlife habitats, since they frequently include native plants that attract a greater variety of birds and small animals. In addition, poplar trees, sugar cane, and other crops can be grown on land unsuitable for food production.[52]

Water Use

Renewables such as wind and solar PV energy do not consume or withdraw water. Also, hydroelectric, geothermal, and biomass facilities do not risk radioactive contamination of water supplies. Studies have shown that renewables can play a key role in averting a "business-as-usual scenario" where "consumption of water in the electric sector could grow substantially,"[53] and that "greater additions of wind to offset fossil, hydropower, and nuclear assets in a generation portfolio will result in a technology that uses no water, offsetting water-dependent technologies."[54]

Three studies assessing population growth, electricity use, and shortages of water during the summer in the US found that 22 counties and 20 large metropolitan areas could experience severe water shortages by 2025, due in large part to thirsty nuclear power plants. The studies noted that nuclear plants could deplete the water available from Lake Lanier in Georgia and exacerbate interstate litigation between Tennessee, Alabama, and Florida. Biodiversity could perish along the Catawba-Wateree River Basin in North Carolina. Chicago could find itself embroiled in domestic and international legal disputes over the consumption and withdrawal of water from Lake Michigan. Households and businesses could run out of water from the South Platte River in Colorado. Rivers could stop recharging the groundwater needed for drinking and irrigation in Texas. Lake Mead and the Colorado River could continue to suffer drought, drastically affecting the state of Nevada and inducing an agricultural crisis in California and Mexico. Fisheries along the Hudson River in New York could collapse. The delta smelt could become extinct in the San Joaquin River Basin in California. These impending but avertable risks serve as an important reminder that climate change is not the only serious environmental issue facing the electricity industry.[55] All three studies noted that, by contrast, energy efficiency measures, commercial wind farms, and residential solar panels could meet all electricity needs in these areas by 2025 without relying on nuclear power or fossil fuels, *and* avoid consuming and withdrawing large amounts of water.

Dr. Ed Brown, Director of Environmental Programs at the University of Northern Iowa, estimated that a 100-W solar panel would save approximately 2,000–3,000 gallons of water over the course of its lifetime. The American Wind Energy Association conducted one of the most comprehensive assessments of renewable energy and water consumption. Their study estimated that wind power uses less than 1/600th as much water per unit of electricity produced as does nuclear power, 1/500th as much as coal power, and 1/250th as much as natural gas. (Small amounts of water are used to clean wind and solar systems.) By displacing centralized fossil fuel and nuclear generation, clean power sources such as energy efficiency and renewables can conserve substantial amounts of water that would otherwise be withdrawn and consumed for the production of electricity.[56]

Environmental Pollution and Climate Change

Every single renewable power technology is less greenhouse gas-intensive than any sized nuclear power plant. A single, 1-MW wind turbine running at only 30% of capacity for one year, for example, displaces more than 1,500 tons of CO_2, 2.5 tons of sulfur dioxide, 3.2 tons of nitrogen oxides, and 60 pounds of toxic mercury emissions.[57] One study assessing the environmental savings of a 580-MW wind farm located at the Altamont Pass near San Francisco, California, concluded that the turbines displace hundreds of thousands of tons of air pollutants each year that would have otherwise resulted from fossil fuel combustion. The study estimated that the wind farm will displace more than 24 billion pounds of nitrogen oxides, sulfur dioxides, particulate matter, and CO_2 over the course of its 20-year lifetime — enough to cover the entire city of Oakland in a pile of toxic pollution 40 stories high.[58]

Dedicated biomass electrical plants release no net CO_2 emissions into the atmosphere (as long as they avoid combusting fossilized fuel) and produce fewer toxic gases. One study conducted by the Imperial Centre for Energy Policy and Technology found that combined-cycle biomass gasification plants produce one-twentieth the amount of pollutants emitted by coal-fired power plants, and one-tenth the pollution of equivalent natural gas plants.[59] Landfill-capture generators and anaerobic digesters harness methane and other noxious gases from landfills and transform them into electricity. This does not just produce useful energy, but also displaces greenhouse gases that would otherwise escape into the environment.

Geothermal plants also have immense air quality benefits. A typical geothermal plant using hot water and steam to generate electricity emits about 1% of the sulfur dioxide, less than 1% of the nitrous oxide, and 5% of the CO_2 emitted by a coal-fired plant of equal size.[60] Its airborne emissions are "essentially nonexistent" because geothermal gases are not released into the atmosphere during normal operation. Another study calculated that the geothermal plants currently in operation throughout the US avoid 32,000 tons of NO_x, 78,000 tons of SO_2, 17,000 tons of particulate matter, and 16 million tons of CO_2 emissions every single year.[61]

All forms of hydroelectric generation combust no fuel, meaning they produce little to no air pollution in comparison with conventional power plants. Luc Gagnon and Joop F. van de Vate conducted a full lifecycle assessment of hydroelectric facilities, and focused on activities related to the building of dams, dykes, and power stations; decaying biomass from flooded land, where plant decomposition produces methane and CO_2; and the thermal backup power needed when seasonal changes cause hydroelectric plants to run at partial capacity. They found that typical emissions of greenhouse gases from hydropower are still 30–60 times less than those from equally sized fossil-fueled stations.[62]

In terms of climate change and greenhouse gases, the International Atomic Energy Agency (IAEA) estimated that, when direct and indirect carbon emissions are included, coal plants are about seven times more carbon-intensive than solar power and 50 times more carbon-intensive than wind technologies. Natural gas fares little better, with two times the carbon intensity of solar power and 27 times the carbon intensity of wind.[63] In the US, the DOE estimated that "every kWh of renewable power avoids the emission of more than one pound of carbon dioxide."[64] According to data compiled by the Union of Concerned Scientists, a think tank, achieving 20% renewables penetration by 2020 would reduce CO_2 emissions by 434 million metric tons — the equivalent of taking nearly 71 million automobiles off the road.[65]

An almost identical study found that biomass facilities are about ten times cleaner than the best coal technologies, and that wind, solar electric, and hydroelectric systems are almost 100 times cleaner than the cleanest coal-fueled system.[66] Martin Pehnt from the Institute for Energy and Environmental Research in Heidelberg conducted lifecycle analyses of 15 separate distributed generation and renewable energy technologies, and found that all but one — solar PV — emitted much less carbon dioxide or other greenhouse gases per kilowatt-hour than nuclear reactors.[67] In an analysis using updated data, researchers from Brookhaven National Laboratory found that current estimates of greenhouse gas emissions for a typical solar PV system range from 29 to 35 grams of CO_2e/kWh, which is significantly less than the equivalent emissions for nuclear power.[68]

The lesson is clear: renewable electricity technologies can successfully replace fossil fuels — we do not need nuclear energy. Nuclear energy proponents may argue that these estimates compare baseload energy sources (such as nuclear) to intermittent or non-dispatchable sources (such as wind and solar PV). However, if these updated numbers are correct, then renewable energy technologies are two to seven times more effective on a per kWh basis than nuclear power at fighting climate change. Therefore, even the deployment of much more intermittent renewable capacity to generate equivalent amounts of energy would still more effectively address climate change than relying on the deployment of baseload nuclear or fossil-fueled generators.

Safety and Security

Unlike the scores of nuclear accidents discussed in Chapter 3, not a single major energy accident in the past century involved small-scale renewable energy systems or energy efficiency; whereas fossil-fueled, nuclear, and larger hydroelectric facilities were responsible for 279 accidents totaling US$41 billion in damages and 182,156 deaths.[69] An investigation of energy-related accidents in the European Union and Russia found that nuclear power was 41 times more dangerous than equivalently sized coal, oil, natural gas, and hydroelectric projects. Nuclear plants were at risk of killing about 46 people for every gigawatt-year of power produced.[70] A database of major industrial accidents from 1969 to 1996 compiled by the Paul Scherrer Institute found that 31% (or 4,290 out of 13,914) were related to the energy sector.[71] Another study concluded that about 25% of the fatalities caused by severe accidents worldwide during the period 1970–1985 occurred in the conventional energy sector.[72] Even if we were to assume that a massive expansion of renewable energy systems may increase the likelihood of industrial accidents within that sector, any reasonable estimate would find that renewables are a far safer alternative to nuclear or fossil fuels.

Deploying renewable power systems in targeted areas also provides an effective alternative to constructing new transmission and distribution (T&D) lines, transformers, local taps, feeders, and switchgears, especially in congested areas or regions where the permitting of new transmission networks is difficult. One study found that up to 10% of total distribution capacity in 10-year high-growth scenarios could be cost-effectively deferred using distributed generation technologies such as solar PV and solar thermal.[73] PG&E, the largest investor-owned utility in California, built an entire power plant in 1993 to test the grid benefits of a 500-kW solar PV plant. PG&E found that the generator improved voltage support, minimized power losses, lowered operating temperatures for transformers on the grid, and improved transmission capacity.[74] The benefits were so large that the small-scale solar PV generator was twice as valuable as the utility had originally estimated, with projected benefits of 14–20 cents/kWh.

The experience convinced PG&E to consider the use of solar PV systems as a substitute for greater investments in T&D infrastructure. Using conventional approaches, planners had proposed an upgrade of high-voltage transmission lines serving seven substations in the San Francisco area, estimated to cost PG&E US$355 million (in 1990 dollars).[75] However, PG&E discovered that a cheaper alternative was to strategically deploy distributed 500-kW solar PV plants connected to distribution feeders. By investing in such locally sited solar PV projects, PG&E found that it could defer a significant number of its transmission upgrades and ultimately saved US$193 million (or more than half the present cost of the expansion plan) by installing solar panels.

Since modern renewable technology enables utilities to remotely dispatch hundreds of scattered units, it also improves the ability of utilities to handle peak load and grid congestion problems. Another PG&E analysis, comparing distributed solar PV plants to one central plant in Kerman, California, found that the grid advantages (in terms of load savings and reduced congestion) more than offset the disadvantages (in terms of high capital cost and interconnection) of installing the new generation.[76]

The use of renewables also diversifies the "fuels" used to generate electricity, thereby minimizing the risk of fuel interruptions, shortages, and accidents. Together, renewable power technologies can increase security by reducing the number of large and vulnerable targets on the grid, providing insulation for the grid in the event of an attack, and minimizing foreign dependence on uranium. Although renewable technologies are constantly derided as intermittent or variable, it is far more certain to rely on the sun shining and the wind blowing than to rely on a system that saboteurs could easily disrupt by blowing up a single power station or snipping a few transmission lines. Renewables are far more resilient and far less attractive a target to possible attackers than the ever-tempting nuclear power plant, spent fuel repository, or uranium mine.

Endnotes

[1] M. King Hubbert, "The Energy Resources of the Earth," in *Energy and Power — A Scientific American Book* (San Francisco: W.H. Freeman and Company, 1971), p. 31.

[2] Quoted in Kate Connolly, "Endless Possibility," *Guardian*, April 16, 2008, p. 9.

[3] Benjamin K. Sovacool and Christopher Cooper, "Nuclear Nonsense: Why Nuclear Power Is No Answer to Climate Change and the World's Post-Kyoto Energy Challenges," *William & Mary Environmental Law and Policy Review* 33(1) (2008), pp. 1–119.

[4] See Charles Komanoff, "Securing Power Through Energy Conservation and Efficiency in New York: Profiting from California's Experience," *Report for the Pace Law School Energy Project and the Natural Resources Defense Council* (May, 2002); Richard Cowart, *Efficient Reliability: The Critical Role of Demand-Side Resources in Power Systems and Markets* (Washington, D.C.: National Association of Regulatory Utility Commissioners, June 2001); and Ahmad Faruqui, Ryan Hledik, Sam Newell, and Hannes Pfeifenberger, "The Power of 5 Percent," *Electricity Journal* 20(8) (October, 2007), pp. 68–77.

[5] David M. Nemtzow, "National Energy Policy: Conservation and Energy Efficiency," in *Hearing Before the House Committee on Energy and Commerce* (June 22, 2001), pp. 76–81.

[6] Martin Schweitzer and Bruce E. Tonn, *An Evaluation of State Energy Program Accomplishments: 2002 Program Year* (Washington, D.C.: US DOE, June 2005, ORNL/CON-492).

[7] International Energy Agency, *The Experience with Energy Efficiency Policies and Programs in IEA Countries: Learning from the Critics* (Paris, France: International Energy Agency, August 2005).

[8] Benjamin K. Sovacool and Anthony D'Agostino, "Nuclear Renaissance: A Flawed Proposition," *Chemical Engineering Progress* 106(7) (July, 2010), pp. 29–35.

[9] Jon Wellinghoff and David L. Morenoff, "Recognizing the Importance of Demand Response: The Second Half of the Wholesale Electric Market Function," *Energy Law Journal* 28(2) (2007), pp. 389–428.

[10] Travis Madsen, Johanna Neumann, and Emily Rusch, *The High Cost of Nuclear Power: Why America Should Choose a Clean Energy Future over New Nuclear Reactors* (Baltimore: Maryland PIRG Foundation, March 2009).

[11] Augusto De La Torre, Pablo Fajnzybler, and John Nash, *Low-Carbon Development: Latin American Responses to Climate Change* (Washington, D.C.: World Bank Group, 2010).

[12] The original figure was US$438 billion in 2000, adjusted to 2007 USD.

[13] Amory Lovins, "Preface to the Chinese Edition of *Winning the Oil Endgame*," February 29, 2008.

[14] Steven Nadel, "National Energy Policy: Conservation and Energy Efficiency," *Hearing Before the House Committee on Energy and Commerce* (June 22, 2001), pp. 46–51.

[15] Fereidoon P. Sioshansi, "The Myths and Facts of Energy Efficiency: Survey of Implementation Issues," *Energy Policy* 19(3) (April, 1991), pp. 231–243.

[16] Ralph Cavanagh, "Restructuring for Sustainability: Toward New Electric Service Industries," *Electricity Journal* 9(6) (July, 1996), p. 72.

[17] Ole Langniss and Barbara Praetorius, "How Much Market Do Market-Based Instruments Create?," *Energy Policy* 34 (2006), pp. 200–211.

[18] Lea Kosnik, "The Potential of Water Power in the Fight Against Global Warming in the US," *Energy Policy* 36 (2008), pp. 3253–3258.

[19] See Benjamin K. Sovacool, "The Intermittency of Wind, Solar, and Renewable Electricity Generators: Technical Barrier or Rhetorical Excuse?," *Utilities Policy* 17(3) (September, 2009), pp. 288–296, for a survey of these articles.

[20] Robert Gross and Philip Heptonstall, "The Costs and Impacts of Intermittency," *Energy Policy* 36 (2008), pp. 4005–4007.

[21] Marilyn A. Brown and Benjamin K. Sovacool, "Promoting a Level Playing Field for Energy Options: Electricity Alternatives and the Case of the Indian Point Energy Center," *Energy Efficiency* 1(1) (February, 2008), pp. 35–48.

[22] William Harvey, "Renewable Energy: Price and Policy Are Key," environmentalresearchweb.org, July 30, 2008.

[23] Paul Denholm, "Improving the Technical, Environmental and Social Performance of Wind Energy Systems Using Biomass-Based Energy Storage," *Renewable Energy* 31 (2006), p. 1356.

[24] Miguel Mendonça, David Jacobs, and Benjamin K. Sovacool, *Powering the Green Economy: The Feed-In Tariff Handbook* (London: Earthscan, 2009).

[25] German Federal Ministry of Economics and Technology, *E-Energy: ICT-Based Energy Systems of the Future* (Berlin: BMWi, April 2008).

[26] United Nations Industrial Development Organization, *UNIDO and Renewable Energy: Greening the Industrial Agenda* (Vienna: UNIDO, 2009), pp. 20–21.

[27] *Ibid.*

[28] US Department of Energy, *PV in Hybrid Power Systems* (Washington, D.C.: DOE Office of Energy Efficiency and Renewable Energy, January 5, 2006), p. 1.

[29] Benjamin K. Sovacool, *The Dirty Energy Dilemma: What's Blocking Clean Power in the United States* (Westport, CT: Praegar, 2008), p. 85.

[30] *Ibid.*

[31] Jeffrey H. Michel, "The Case for Renewable FITs," *Journal of EUEC* 1 (2007), pp. 2–19.

[32] Benjamin K. Sovacool and Richard F. Hirsh, "Beyond Batteries: An Examination of the Benefits and Barriers to Plug-in Hybrid Electric Vehicles (PHEVs) and a Vehicle-to-Grid (V2G) Transition," *Energy Policy* 37(3) (March, 2009), p. 1096.

[33] PSE poured the first foundation on May 18, 2005, and the Hopkins Ridge Wind Project began commercial operations on November 27, 2005. See Roger Garratt, Director, Resource Acquisition & Emerging Technologies, "Puget Sound Energy — Exploring Wind & Solar Resources," Presentation at the "Harvesting Clean Energy" Conference (January 29, 2008), p. 6.

[34] Ormat Technologies, Inc., "ORMAT's State of the Art Geothermal Power Plant, Commissioned Eight Months After Ground Breaking," Press Release (November 15, 2005).

[35] Partnership for Advancing Technology in Housing, *Harnessing the Sun: Passive and Active Solar Systems Offer Growing Niche Market* (2006), p. 4.

[36] Amory Lovins *et al.*, *Small Is Profitable: The Hidden Benefits of Making Electrical Resources the Right Size* (Snowmass, CO: Rocky Mountain Institute, 2002), p. 132.

[37] *Ibid.*, p. 252.

[38] Amory B. Lovins and L. Hunter Lovins, *Brittle Power: Energy Strategy for National Security* (Andover, MA: Brick House Publishing Company, 1982).

[39] The United Nations reported these findings in Renewable Energy Policy Network for the 21st Century, *Renewables 2007: Global Status Report* (Washington, D.C.: REN21, 2008).

[40] Renewable energy figures taken from Renewable Energy Policy Network for the 21st Century (2008). Nuclear power figures taken from Pam Radtke Russell, "Prices Are Rising: Nuclear Cost Estimates Under Pressure," *EnergyBiz Insider* (May–June, 2008), p. 22.

[41] Madsen *et al.* (2009).

[42] Michael Replogle, "Overcoming Barriers to Transportation Cost Internalization," in Olav Hohmeyer, Richard L. Ottinger, and Klaus Rennings (eds.), *Social Costs and Sustainability: Valuation and Implementation in the Energy and Transport Sector* (New York: Springer, 1997), pp. 433–447.

[43] Sovacool (2008), pp. 109–110.

[44] Tom Buis, "Concerning the Renewable Energy Economy: A New Path to Investment, Jobs, and Growth," Testimony before the House Select Committee on Energy Independence and Global Warming, March 6, 2008.

[45] US Department of Energy, *Characterization of U.S. Energy Resources and Reserves* (1989), p. 19.

[46] National Research Council, *Electricity from Renewable Resources: Status, Prospects, and Impediments* (Washington, D.C.: The National Academies Press, 2009), pp. 2–3.

[47] Electricity generation taken from US Energy Information Administration, *Electricity Net Generation from Renewable Energy by Energy Use Sector and Energy Source, 2002–2006* (Washington, D.C.: US DOE, 2007). Achievable onshore wind potential assumes class 1–7 wind regimes in all 50 states (and is

based on the DOE estimate that onshore wind could supply "more than one and a half times the current electricity consumption of the U.S."). See Energy Efficiency and Renewable Energy Program of the US Department of Energy, *Wind Energy Resource Potential* (Washington, D.C.: DOE, 2007). Achievable offshore wind potential assumes water depths from zero to 900 m. The estimate excludes 266,200 MW of offshore potential for waters currently deeper than 900 m because such technology is not commercially available. Data taken from Walt Musial, *Offshore Wind Energy Potential for the U.S.* (Golden, CO: NREL, May 19, 2005), p. 9. Achievable solar PV potential assumes prices of US$2–$2.50 per installed watt. Data taken from Maya Chaudhari, Lisa Frantzis, and Tom E. Hoff, *PV Grid Connected Market Potential* (San Francisco: The Energy Foundation, September 2004). Achievable solar thermal potential includes parabolic troughs and power towers, and is taken from National Renewable Energy Laboratory, *Concentrating Solar Power Resource Maps* (Golden, CO: NREL, December 2007). The NREL states that "realistically, the potential of concentrating solar power in the Southwest could reach hundreds of gigawatts or greater than 10% of U.S. electric supply." Achievable geothermal potential taken from Bruce D. Green and R. Gerald Nix, *Geothermal — The Energy Under Our Feet* (Golden, CO: NREL, November 2006, NREL/TP-840-40665). Achievable biomass potential (combustion) converted from estimates provided in Oak Ridge National Laboratory and US Department of Energy, *Biomass as Feedstock for a Bioenergy and Bioproducts Industry: The Technical Feasibility of a Billion-Ton Annual Supply* (Washington, D.C.: US DOE, 2005, DOE/GO-102995-2135). Achievable biomass potential (landfill gas) taken from US Environmental Protection Agency, *An Overview of Landfill Gas Energy in the U.S.* (Washington, D.C.: Landfill Methane Outreach Program, May 2007). Achievable hydroelectric potential excludes all nationally protected lands and areas, and is taken from US Department of Energy, *Water Resources of the U.S. with Emphasis on Low Head/Low Power Resources* (Washington, D.C.: DOE, April 2004, DOE/ID-11111).

[48] Most estimates assume the exclusion of low-resource and environmentally sensitive areas. Some figures for concentrated solar power and geothermal are derived from the mean of the given ranges presented. All figures are taken from Mark Z. Jacobson, "Review of Solutions to Global Warming, Air Pollution, and Energy Security," *Energy & Environmental Science* 2 (2009), pp. 148–173, except

for the electrical potential of hydroelectric power, which is taken from International Hydropower Association, *Hydropower and the World's Energy Future* (Bonn: IHA, November 2000).

[49] Sovacool and Cooper (2008).

[50] National Renewable Energy Laboratory, *PV FAQs* (Golden, CO: NREL, 2007).

[51] *Ibid.*

[52] *Ibid.*

[53] US Department of Energy, *Energy Demands on Water Resources: Report to Congress on the Interdependency of Energy and Water* (2006), pp. 10–11.

[54] US Department of Energy, *The Wind/Water Nexus* (2006), p. 2.

[55] See Benjamin K. Sovacool, "Running on Empty: The Electricity–Water Nexus and the U.S. Electric Utility Sector," *Energy Law Journal* 30(1) (April, 2009), pp. 11–51; Benjamin K. Sovacool and Kelly E. Sovacool, "Identifying Future Electricity–Water Tradeoffs in the United States," *Energy Policy* 37(7) (July, 2009), pp. 2763–2773; and Benjamin K. Sovacool and Kelly E. Sovacool, "Preventing National Electricity–Water Crisis Areas in the United States," *Columbia Journal of Environmental Law* 34(2) (July, 2009), pp. 333–393.

[56] Sovacool and Cooper (2008).

[57] Ari Reeves, *Wind Energy For Electric Power: A REPP Issue Brief* (Washington, D.C.: Renewable Energy Policy Project, 2003), p. 4.

[58] PowerWorks, *Health and Climate Benefits of Altamont Pass Wind Power* (September 29, 2008).

[59] Ausilio Bauen, Jeremy Woods, and Rebecca Hailes, *Bioelectricity Vision: Achieving 15% of Electricity from Biomass in OECD Countries by 2020* (London: ICEPT, 2004), p. 25.

[60] Wendell A. Duffield and John H. Sass, *Geothermal Energy — Clean Power from the Earth's Heat* (Reston, VA: USGS, 2003), p. 26.

[61] Alyssa Kagel and Karl Gawell, "Promoting Geothermal Energy: Air Emissions Comparison and Externality Analysis," *Electricity Journal* (August/September, 2005), p. 92.

[62] Luc Gagnon and Joop F. van de Vate, "Greenhouse Gas Emissions from Hydropower: The State of Research in 1996," *Energy Policy* 25 (1997), p. 8.

[63] See Joseph V. Spadaro, Lucille Langlois, and Bruce Hamilton, "Greenhouse Gas Emissions of Electricity Generation Chains: Assessing the Difference," *IAEA Bulletin* (March, 2000), pp. 19–21.

[64] US Department of Energy *et al.*, *Guide to Purchasing Green Power: Renewable Electricity, Renewable Energy Certificates, and Onsite Renewable Generation* (2004), p. 2.

[65] Union of Concerned Scientists, *Successful Strategies: Renewable Electricity Standards* (2008), p. 2.

[66] Luc Gagnon, Camille Belanger, and Yohji Uchiyama, "Life-Cycle Assessment of Electricity Generation Options: The Status of Research in Year 2001," *Energy Policy* 30 (2002), p. 1271.

[67] See Martin Pehnt, "Dynamic Life Cycle Assessment of Renewable Energy Technologies," *Renewable Energy* 31 (2006), p. 60.

[68] Vasilis M. Fthenakis, Hyung Chul Kim, and Erik Alsema, "Emissions from Photovoltaic Life Cycles," *Environmental Science & Technology* 42 (2008), p. 2170.

[69] Benjamin K. Sovacool, "The Costs of Failure: A Preliminary Assessment of Major Energy Accidents, 1907–2007," *Energy Policy* 36 (2008), p. 1807.

[70] Stefan Hirschberg and Andrzej Strupczewski, "How Acceptable? Comparison of Accident Risks in Different Energy Systems," *IAEA Bulletin* (March, 1999), pp. 25–27.

[71] Stefan Hirschberg *et al.*, "Severe Accidents in the Energy Sector: Comparative Perspective," *Journal of Hazardous Materials* 111 (2004), p. 58.

[72] Andrew F. Fritzsche, "Severe Accidents: Can They Occur Only in the Nuclear Production of Electricity?," *Risk Analysis* 12 (1992), p. 327.

[73] R.G. Pratt *et al.*, "Potential for Feeder Equipment Upgrade Deferrals in a Distributed Utility," Presentation at the American Council for an Energy Efficient Economy 1994 Summer Conference (1994), p. 2.

[74] Howard J. Wenger, Thomas E. Hoff, and Brian K. Farmer, "Measuring the Value of Distributed Photovoltaic Generation: Final Results of the Kerman Grid-Support Project," in *1994 IEEE First World Conference on Photovoltaic Energy Conversion, Volume 1* (1994), p. 793.

[75] Charles D. Feinstein, Ren Orans, and Stephen W. Chapel, "The Distributed Utility: A New Electric Utility Planning and Pricing Paradigm," *Annual Review of Energy and the Environment* 22 (1997), pp. 159–160.

[76] T. Hoff and D.S. Shugar, "The Value of Grid-Support Photovoltaics in Reducing Distribution System Losses," *IEEE Transactions on Energy Conversion* 10 (1995), pp. 569–576.

8

The "Self-Limiting" Future of Nuclear Power

Benjamin Franklin once wrote that "the great advantage of being a reasonable creature is that you can find a reason for whatever you do." The nuclear power industry possesses no shortage of arguments in favor of a nuclear renaissance, many of them reasonable at first glance. Yet, the central premise of this book is that a global nuclear renaissance would bring immense technical, economic, environmental, political, and social costs. Nuclear power generators cannot be mass-produced. As Table 1 summarizes, they take much longer to build, and are therefore exposed to escalating interest rates, inaccurate demand forecasts, and unforeseen labor conflicts. Their centralization requires costly and expansive transmission and distribution systems. Modern nuclear reactors are prone to a deteriorating energy payback ratio for the nuclear fuel cycle, produce hazardous and extremely long-lived waste, have large water requirements, and possess a larger carbon footprint than energy efficiency and every form of renewable electricity.

All is not lost, however. As this book has also shown, renewable power technologies reduce dependence on foreign sources of uranium, and therefore create a more secure fuel supply chain that minimizes exposure to economic and political changes abroad. Renewable technologies decentralize electricity supply, so that an accidental or intentional outage would

Table 1:   **Disadvantages of Modern Nuclear Power Plants**

Dimension	Category	Explanation
Technical	Safety and accidents	Human error and technological failure have resulted in hundreds of incidents and accidents; the impact of a serious accident, even if the probability is low, would be catastrophic
	Materials and labor	A shortage of key components and skilled labor could result in increased costs for nuclear power plants and/or slow any transition to a nuclear renaissance
	Fuel availability and energy payback	The energy intensity of the nuclear fuel cycle and declining reserves of high-quality uranium result in a low energy payback ratio, whereby plants must operate for decades before they produce any net energy
Economic	Construction and operating costs	Long construction lead times for new plants create substantial risk of cost overruns and create expected high future operating costs
	Reprocessing costs	Reprocessing of nuclear fuel costs billions of dollars and creates its own security risks related to the availability of plutonium
	Waste storage costs	Hundreds of millions of dollars each year must be spent on onsite storage, to say nothing of the gargantuan cost of building permanent geologic repositories for spent fuel
	Decommissioning costs	Decommissioning costs can sometimes be greater than the costs of building a plant in the first place
	Fuel costs	Uranium reserves are consolidated among a small number of countries, and dependence on foreign suppliers runs the risk of disruption and price volatility
	Security costs	Nuclear facilities must be rigorously guarded and protected
	Research costs	The next generation of nuclear reactors will require billions of dollars in research funds and subsidies
Environmental	Land use — uranium mining	Underground mining, open-pit mining, and *in situ* leaching of uranium create serious environmental hazards and can contaminate water supplies

(Continued)

Table 1:   (*Continued*)

Dimension	Category	Explanation
	Land use — waste storage	More than 10,000 metric tons of waste are created each year by commercial nuclear reactors around the world; the waste they produce is extremely hazardous and difficult to store
	Water use and contamination	Existing nuclear power plants consume and withdraw vast quantities of water needed for operation; risk entrainment, impingement, and thermal discharges; and can contaminate water supplies with tritium and other radioactive pollutants
	Climate change	The carbon footprint for a typical nuclear reactor could be equivalent to that of fossil fuels in the next few decades if high-grade uranium ores continue to be exhausted, as nuclear reactors entail considerable greenhouse gas emissions from their lifecycle (much greater than from renewable energy resources and some other alternatives); the heat discharges from nuclear power plants also indirectly contribute to global warming
	Medical and health risks	Operating nuclear reactors have been shown to create health risks for local communities and workers
Sociopolitical	Transmission and distribution vulnerability	Nuclear power plants rely on a complex distribution system that is subject to cascading failures easily induced by severe weather, human error, sabotage, or even the interference of small animals
	Plant and reactor insecurity	Nuclear power plants and research reactors continue to be attractive targets for terrorists and criminals
	Weapons proliferation	All stages of the nuclear fuel cycle produce fissile material that can be used to manufacture weapons of mass destruction
	Military conflict	Nuclear power plants are often bombed and attacked during military campaigns
	Maritime and transport security	The movement of nuclear fuel and waste is subject to accidents, piracy, and theft
	Community marginalization	Nuclear facilities are often sited and located in peripheral areas that marginalize communities

affect a smaller amount of capacity than an outage at a larger nuclear facility. Renewable energy technologies improve the reliability of power generation by conserving or producing power close to the end-user, and by minimizing the need to produce, transport, and store hazardous and radioactive fuel. Unlike generators relying on uranium and recycled plutonium, renewable generators are not subject to the volatility of global fuel markets. They can also respond more rapidly to supply and demand fluctuations, improving the efficiency of the electricity market. Most significantly, renewable power technologies have enormous environmental benefits, since their use tends to avoid air pollution and the dangers and risks of extracting uranium. They generate electricity without releasing significant quantities of CO_2 and other greenhouse gases that contribute to climate change as well as life-endangering nitrogen oxides, sulfur dioxides, particulate matter, and mercury. They also create power without relying on the extraction of uranium and its associated digging, drilling, mining, leaching, transporting, storing, sequestering, and polluting of land.

In the end, nuclear reactors and renewable power generators do the same thing: they produce electrical energy (kWh). Why rely on a nuclear system that is subject to highly uncertain projections about uranium availability, centrally administered by technocratic elites, and vulnerable to the ebb and flow of international politics (requiring garrison-like security measures at multiple points in the supply chain), when superior alternatives exist?

The simple fact that energy efficiency programs and renewable power technologies are better than nuclear power plants has not been advanced by this book alone. Indeed, consider these following studies, all of which reach a similar conclusion:

- "The limited prospects for nuclear power today are attributable, ultimately, by four unresolved problems: high relative costs; perceived adverse safety, environmental, and health effects; potential security risks stemming from proliferation; and unresolved challenges in long-term management of nuclear wastes"[1];
- "Because of hasty commercialization, safety concerns, and unresolved long-term storage of its wastes, the first nuclear era has been a peculiarly successful failure"[2];

- "The economics are profoundly unfavorable and are getting worse. A significant expansion of nuclear energy worldwide to 2030 faces constraints that ... are likely to outweigh the drivers of nuclear energy"[3];
- "The failure of the U.S. nuclear power program ranks as the largest managerial disaster in business history, a disaster on a monumental scale.... [O]nly the blind, or the biased, can now think that the money has been well spent. It is a defeat for the U.S. consumer and for the competitiveness of U.S. industry, for the utilities that undertook the program and for the private enterprise system that made it possible"[4];
- "There is no convincing case for building new nuclear power plants anywhere in the world for the sake of business. Nor is there a convincing case for updating or rehabilitating existing plants, especially in light of the limited availability of useful fissile material, considerable risks involved at all stages of production, very high legacy costs, imposed on thousands of future generations, absence of secure long-term waste storage, and additional risks and wastes resulting from reprocessing"[5];
- "No other energy technology spreads do-it-yourself kits and innocent disguises for making weapons of mass destruction, nor creates terrorist targets or potential for mishaps that can devastate a region, nor creates wastes so hazardous, nor is unable to restart for days after an unexpected shutdown"[6];
- "The accumulated experience of the past six decades provides ample evidence of adverse health effects in workers in the nuclear fuel cycle, the potential for disastrous accidents that lead to widespread environmental contamination, the unresolved problems of permanent and secure storage of high-level radioactive wastes, and the extraordinarily high costs of building additional nuclear power generation facilities.... Given the availability of alternative carbon-free and low-carbon options and the potential to develop more efficient renewable technologies, it seems evident that public health would be better served in the long term by these alternatives than by increasing the number of nuclear power plants"[7]; and
- "We may not need any [new nuclear power plants,] ever.... Renewables like wind, solar, and biomass will provide enough energy to meet baseload capacity and future demands."[8]

Who has made such claims? The sources are, in order, an interdisciplinary team from the Massachusetts Institute of Technology, the historian and energy analyst Vaclav Smil, an independent study from the Centre for International Governance Innovation, *Forbes Magazine*, a recent 2010 dialogue on the future of nuclear energy, the physicist Amory Lovins, a physician writing in *Environmental Health Perspectives*, and Federal Energy Regulatory Commission Chairman David Wellinghoff. The fact that these quotes come from a variety of sources (academic journals, magazines, and reports) across the political spectrum (including business, science, civil society, and medicine) and from different disciplines (physics, economics, epidemiology, and politics) suggests that there is a consensus among a broad base of independent, nonpartisan experts that nuclear power plants are a poor choice for producing electricity.

So why, then, does nuclear power persist? One study supposed that it is the superficially attractive narrative associated with nuclear energy that conflates it with national progress and pride, alongside an immensely powerful and effective lobby, a new generation that has either forgotten or never known why it failed previously, deeply rooted habits that favor giant power stations, and lazy reporting by a credulous press.[9] This chapter argues that three primary culprits exist: the true costs of nuclear energy are not borne by those benefiting from it, resulting in what economists call "market failure"; many of the costs and risks involved with nuclear electricity are passed directly onto ratepayers; and nuclear power has, since its inception, been associated with complex notions of progress and modernity that make it seductive, despite all of its intractable challenges. Taken together, these three culprits — market failure and externalities, the socialization of risk, and hubris and technological fantasy — largely explain why nuclear power plants flourish. When these conditions change (i.e. when the full costs of nuclear energy become apparent or can no longer be socialized, or when the allure of nuclear fission fades), the drive towards nuclear energy stalls. In short, if nuclear energy is to have any future at all, it will be what Joseph Romm has called a "self-limiting" one.[10]

Market Failure and Externalities

As almost any smart undergraduate student of economics knows, free markets for anything — from tomatoes to Tomahawk missiles — need

multiple criteria to function properly. One of them is that all costs must be
fully internalized in the price of a given good or commodity; if one person
is able to shift the costs to someone else while still reaping the benefits,
then the market has failed to distribute benefits equally and equitably,
creating what is known as market failure. At the heart of the market
failure discussion is the concept of an externality.

Defined as costs and benefits resulting from an activity that do not
accrue to the parties involved in the activity, externalities have won atten-
tion in recent decades as an important (albeit often ignored) aspect of
energy production and use.[11] Externalities are part of the "overall social
cost of producing energy . . . including the value of any damages to the
environment, human health, or infrastructure."[12] Another definition of
externalities is "inadvertent and unaccounted for effects of one or more
parties on the welfare of another."[13]

Take the classic example of unregulated pollution from a smokestack.
A factory produces items that are priced by taking into account the
demand for the products as well as labor, capital, and other costs, but the
damages from the factory's pollution — health and other effects — are
true costs borne by society that are unaccounted for in the price of the
factory's product. These latter costs are commonly referred to as "externalities" because people tend to consume them as byproducts of other
activities that are external to market transactions and, therefore, unpriced.
This means that the factory produces a volume of items that is less than
"socially optimal," resulting in a net welfare loss to society in the form of
morbidity, mortality, and reduced productivity.

Nuclear power plants have a plethora of these types of externalities
that most producers and users of nuclear energy do not have to pay for.
A partial list would at least include:

- Catastrophic risks such as nuclear meltdowns and accidents;
- An increased probability of wars due to rapid uranium extraction,
  the boom and bust cycles of uranium mining communities, or the
  inability to secure fissile materials associated with the nuclear fuel
  cycle;
- Public health issues such as chronic exposure to radiation and its consequent advanced morbidity and mortality, as well as worker exposure
  to toxic substances and occupational accidents and hazards;

- Direct land use by power plants, uranium mines, enrichment stations, and storage facilities;
- The destruction of land by uranium mining and leaching, including acid drainage and resettlement;
- The effects of water pollution on fisheries and freshwater ecosystems, which are sensitive to water chemistry, as well as the release of radionuclides into water sources;
- Consumptive water use, with consequent impacts on agriculture and ecosystems where water is scarce;
- Continual maintenance of caches of spent nuclear fuel;
- Changes to the local and regional economic structure through the loss of labor and jobs, transfer of wealth, and reductions in gross domestic product; and
- Incidence of noise and reduced amenity, lower property values near nuclear plants, and aesthetic objections.

Even though this list is incomplete, one study analyzed 132 externality estimates associated with electricity generation in a variety of countries with an assortment of different energy systems.[14] The study found that net social costs for nuclear power ranged from a low of less than 1 cent per kWh to a high of almost 65 cents per kWh, with a mean of 8.6 cents per kWh. As Table 2 documents, the external costs for nuclear power were twice as high as that of hydroelectric systems, more than 12 times higher than that of solar power, and almost 30 times higher than that of wind power. The amount of 8.6 cents per kWh may not sound like much; but if correct, it means that, since nuclear units produced 2,601 billion kWh of

Table 2: **Negative Externalities Associated with Nuclear and Renewable Sources of Electricity (cents/kWh, in 1998 USD)**

	Nuclear	Biomass	Hydroelectric	Solar	Wind
Minimum	0.0003	0	0.02	0	0
Maximum	64.45	22.09	26.26	1.69	0.80
Mean	8.63	5.20	3.84	0.69	0.29
Standard deviation	18.62	6.11	8.40	0.57	0.20

energy in 2008, they also generated US$223.7 billion in global social and environmental damages.

In other words, nuclear power generation created US$223.7 billion of additional costs that are not assumed in traditional estimates of nuclear power's price. Many of these costs are "hidden" because neither nuclear producers nor consumers have to pay for these additional expenses. Instead, the external costs of nuclear energy are shifted to society at large. What is interesting is that — when one takes the negative externalities associated with nuclear power, fossil fuels, and renewable sources of electricity, and adds them on top of existing production costs — Figure 1 shows that wind, geothermal, hydroelectric, and biomass plants are already cheaper than existing nuclear units. Put simply, if the true cost of nuclear energy matched its price, nuclear energy would never be competitive with renewable energy (or energy efficiency) in any free market.

Subsidies and the Socialization of Risk

Because of their capital intensity and financial risk, nuclear power plants are only cost-competitive when they are underwritten with gargantuan public subsidies. Put in other terms, absent an enormous diversion of taxpayer funding, no rational investor would ever finance a nuclear power plant. As one economist put it, investing in nuclear power without the provision of government subsidies is about as useful as "watching a movie with the sound turned off."[15] One 2009 assessment of the global nuclear industry identified no less than ten types of subsidies given to nuclear power plant operators around the world, as presented in Table 3.

Consider the US, where one would think that the electricity market operated freely and with little distortion from subsidies. In fact, the US electricity sector is heavily subsidized, and most subsidies have gone to nuclear power plants. From 1947 to 1999, federal subsidies for nuclear power in the US totaled US$145.4 billion (in 1999 USD). Even in fiscal year 1979, when subsidies for renewable energy peaked in the US at US$1.5 billion, the Department of Energy (DOE) devoted more than 58% of its research budget to nuclear power.

The Energy Policy Act of 1992 promised US$100 million in new funding for reactor designs, set limits on utility payments for decommissioning,

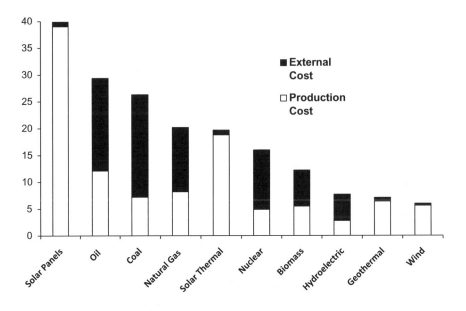

Technology	Market Cost	External Cost
Solar Panels	39	0.9
Oil	12.1	17.3
Coal	7.2	19.1
Natural gas	8.2	12
Solar thermal	18.8	0.9
Nuclear	4.9	11.1
Biomass	5.5	6.7
Hydroelectric	2.8	4.9
Geothermal	6.4	0.7
Wind	5.6	0.4

Figure 1: Production and External Costs for Electricity Generators (in US cents/kWh)

and delegated the authority to set waste disposal standards to the National Academy of Science rather than public participation. However, it failed to incentivize *anyone* to build a new nuclear power plant.[17] The Energy Policy Act of 2005 only worsened the disparity by lavishing the nuclear industry with US$13 billion worth of loan guarantees, US$3 billion in research, US$2 billion in public insurance against delays, US$1.3 billion in tax

Table 3:   Subsidies Common to Nuclear Power Plants Around the World[16]

Type	Subsidy	Explanation	Examples
Capital costs	Subsidized access to credit	Policies that dramatically reduce the cost of capital for nuclear plants by enabling them to obtain debt at the government's cost of borrowing, and to use high levels of this inexpensive debt rather than much more expensive equity	Direct government loans Government guaranteed loans Direct government investment in nuclear-related infrastructure
	Rate-basing of in-process plants	Policies that allow recovery of plant investment prior to commencing operations, and that shift performance and investment risks from owners to ratepayers	Work-in-process allowance for funds used during construction
	Subsidized capital goods	Policies that reduce the after-tax cost of capital goods deployed in the nuclear sector; in the case of R&D, the internal cost to develop new product lines or modify old ones is reduced	Accelerated depreciation Research and development Investment tax or production tax credits Capital write-offs transferred to taxpayer
Operating costs	Fuel and enrichment	Policies that socialize the risks of building, operating, and remediating fuel chain facilities, and that reduce the cost of fuel inputs to reactors	Government-owned or government-subsidized enrichment facilities Subsidized access to uranium ore
	Accident and attack risks	Policies that reduce insurance costs for all participants of the nuclear fuel chain, and that shift accident risks from investors to the surrounding population and taxpayers	Caps on mandated liability coverage
	Industry oversight	Subsidies that disadvantage less oversight-intensive competitors, if not fully funded by user fees	Government oversight of domestic industry International oversight through IAEA
	Emissions	Windfall grants of carbon credits that can be immediately resold, and earmarked funds	Privileges under carbon constraints

(Continued)

Table 3: *(Continued)*

Type	Subsidy	Explanation	Examples
Waste management, plant closure	Nuclear waste management	Policies that convert this very high-risk, capital-intensive, fixed-cost endeavor into something the reactors (and investors) no longer have to worry much about	Government-run long-term management of reactor waste
			Payments to existing reactors to store waste onsite
	Plant decommissioning, remediation	Policies that reduce the break-even charges needed for nuclear operations; for fuel chain facilities, very large public liabilities result	Tax-advantaged accrual of decommissioning funds
			Government-provided decommissioning support
Market price support	Market onuses and incentives	Policies that enable nuclear plants to earn higher revenues on power sales than they would be able to in a competitive market	Inclusion of nuclear power in renewable energy portfolios or feed-in tariffs
			Transfer of capital costs to ratepayers via stranded cost rules, or similar transfer of cost recovery

breaks, an extra 1.8 cents/kWh in operating subsidies, and limited liability for accidents. Yet even this was not enough, despite the fact that these subsidies covered 80% of the costs of a new nuclear plant.[18]

These subsidies are in addition to numerous other benefits the nuclear industry already enjoys: free offsite security, no substantive public participation or judicial review of licensing, and payments to operators to store waste. The subsidy established by the Price–Anderson Act, which ironically charges taxpayers for liability insurance against nuclear accidents that could kill them, alone is possibly estimated to be worth more than twice the entire research budget of the US DOE.[19] According to one estimate, nuclear power operators would be responsible for only 2% of the cost of a worst-case accident, with taxpayers picking up the rest of the tab.[20] Interestingly, this very issue of limited liability for nuclear plants could derail the recent "123 deal" made between the US and India. For the deal to go through, Indian legislation must cap nuclear liability; but when lawmakers put forth a Civil Nuclear Liability Bill that limited damages at US$450 million in the event of a nuclear accident, the Indian Supreme Court argued that it violated Article 21 of the Indian Constitution. The presiding judge in the case stated that the main lesson from the Bhopal disaster was that foreign hazardous industries must be made absolutely liable for any damage caused from their facilities.[21]

One interesting comparison is to look at subsidies for wind, solar, and nuclear power for their respective first 15 years of operation. Nuclear power in the US received subsidies worth US$15.30 per kWh between 1947 and 1961, compared to subsidies worth only US$7.19 per kWh for solar power and 46 cents per kWh for wind power between 1975 and 1989. During the first 15 years, nuclear and wind power produced about the same amount of energy: 2.6 billion kWh for nuclear power, and 1.9 billion kWh for wind power. But, nuclear subsidies outweighed wind subsidies by more than a factor of 40, receiving US$39.4 billion compared to wind's US$900 million over the 15-year period.[22]

The trend of grossly subsidizing nuclear energy holds true globally, as nuclear power has received more public research funding than any other source since the 1970s.[23] This is especially true for many other industrialized countries, including Canada, France, Germany, Japan, Sweden, and the UK (illustrated in Figure 2). As the numbers show, nuclear energy has

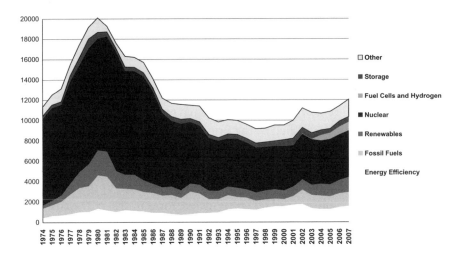

	1974–2007		1998–2007	
Group	Cumulative Total	% Share	Cumulative Total	% Share
Energy efficiency	38,422	8.9	14,893	14.2
Fossil fuels	55,027	12.8	11,114	10.6
Renewable energy	37,333	8.7	10,709	10.2
Nuclear fission and fusion	236,328	54.8	43,667	41.5
Hydrogen and fuel cells	2,824	0.7	2,824	2.7
Transmission and storage	15,717	3.6	5,388	5.1
Other	45,204	10.5	16,599	15.8
Total	**430,855**	**100**	**105,194**	**100**
Nuclear Share of Country Total (%)				
Canada		39.0		28.8
France		81.4		72.5
Germany		67.0		41.0
Japan		72.7		67.2
Sweden		15.2		6.7
United Kingdom		69.0		32.7
United States		38.1		13.2

Figure 2: Government-Funded Subsidies for Nuclear Fission and Fusion Within International Energy Agency Countries, 1974–2007 (in millions of 2007 USD)

received 54.8% of *all* research subsidies among International Energy Agency (IEA) countries, compared to only 8.7% for renewables and 8.9% for energy efficiency.

It may come as no surprise that the only way for utilities to embrace new nuclear units is to receive large subsidies or raise electricity prices for consumers. Some states now allow utilities to increase electricity rates to finance new plants years before construction even begins.[24] In Georgia, these rate increases will amount to a "subsidy" of US$14 billion on top of an additional US$8.3 billion of federal loan guarantees given by the Obama administration.[25] In Levy County, Florida, residential customers will begin paying US$100 per year in higher bills from 2009 to 2016 to help Progress Energy fund a new nuclear unit. South Carolina had to pass a 37% rate hike before it could consider financing a new reactor.[26]

How does the nuclear industry get such sweet subsidies? Part of the explanation may lie in lobbying. In the US, the Investigative Reporting Workshop at American University found that the nuclear industry spent more than US$600 million on lobbying and US$63 million on campaign contributions from 1999 to 2009.[27] In many ways, the nuclear power industry's efforts to win support are a textbook case of how the influence game is played in Washington. Besides the money spent on lobbying and campaign contributions, the industry — led by the Nuclear Energy Institute (NEI) — has created a network of allies who give speeches, quote one another approvingly, and showcase one another on their websites. The effect is an echo chamber of support for nuclear power.

Hubris and Technological Fantasy

One final factor pushing nuclear power is its association with progress, complexity, and modernity. Early advocates promised not only a future of electricity too cheap to meter, but an age of peace and plenty (without high prices or shortages) in which atomic energy would provide the power needed to desalinate water for the thirsty, irrigate deserts for the hungry, and fuel interstellar travel deep into outer space. Other exciting opportunities included atomic golf balls that could always be found and a nuclear-powered airplane, which the US federal government even spent US$1.5 billion researching between 1946 and 1961.[28]

This section suggests that one explanation for the attractiveness of nuclear energy could be its association with national visions of progress. While these visions vary by country and over time, John Byrne and Steven Hoffman propose that the single most consistent predictor of whether a society will embrace nuclear energy is their ability to think in the "future tense." That is, planners and promoters become enthralled by the possible benefits of nuclear energy in the *future*, and are willing to accept the costs in the *present* to realize them. Put another way, they tend to overestimate the advantages of nuclear energy and discount its future costs in the absence of knowledge about current economic or technical compatibility; the reality of present risks and costs is discounted by the unrealized possibilities of future gain.[29] Indeed, the energy historian Martin Melosi has noted that "it's amazing that commercialization of nuclear power occurred at all. . . . The energy market had little to do with this important event, since there was no pressing need for a new source of power in the United States. There was, however, strong interest in enhancing American prestige."[30] Although these psychological benefits are intangible, they are often believed to be real. A cursory look at the genesis of nuclear programs in eight countries — China, France, India, Japan, the former Soviet Union, the US, Spain, and Canada — reveals that, in each case, optimism in the technology and an overarching vision of what nuclear energy could deliver in the future played a role in trumping concerns about present costs.

China, 1953–1992

The prospect of developing nuclear power was first broached in China's first Five-Year Plan in 1953, which emphasized the need for a centralized nuclear development program managed by the government and state enterprises. China's commercial nuclear program formally began in 1972, when the central government approved the first nuclear program — known as the 728 Project — to develop submarine reactors. Nuclear energy quickly became attached to aspirations of Chinese economic power and the legitimatization of China as a superpower.

Throughout the 1970s and 1980s, China experienced massive deficits in electricity supply, with annual demand for electricity surpassing supply by as much as 70 billion kWh. The government had to replace more than 100,000 boilers at conventional power plants between 1972 and 1978, and rolling blackouts hit every major province within China at least twice a year for much of the two decades. Nuclear power was seen as instrumental in overcoming the energy supply deficits, improving Chinese economic competitiveness, "catching up" with Taiwan and other industrialized countries, and enhancing national prestige. Chinese officials even toyed with the idea of exporting both nuclear technology and electricity to the rest of Asia; and built one facility, the Yibin Fuel Component Factory in Sichuan, to manufacture prefabricated components of nuclear power plants for export. They sold one set of components to Pakistan in 1989, and planned to earn billions of dollars of foreign exchange exporting similar packages to Africa and the rest of the developing world.[31]

France, 1945–1970

Left in the devastation caused by the German occupation and fighting of 1944–1945, French technical and scientific experts linked nuclear power to French "radiance" and identity.[32] Nuclear energy was central to this campaign of French economic modernization; and research, development, and construction were dominated by the government. The *Commissariat à l'énergie atomique* (CEA), formed in 1945, had a close association with the bureaucracy in Paris and the military, and was charged with developing indigenous French reactors.[33]

Nuclear energy was seen as a tool to not only provide much-needed electricity to France, but also revitalize the national economy. Nuclear reactors offered the chance for French planners to rebuild infrastructure, promote industry, and augment political influence simultaneously. One key component of this push was the notion of *dirigisme*, or the idea that government-led intervention and planning was the best way to respond to social problems. Another component was the notion of French "national champions," or the idea that key sectors of the economy (such as the state-owned nuclear manufacturer Framatome) deserved special protection and support from the government.[34] After the creation and demonstration of the atomic bomb, "nuclear technology became a quintessential symbol of modernity and national power."[35]

India, 1945–1980

The Indian government began investigating nuclear energy in 1945, when they formed the Tata Institute of Fundamental Research and appointed a prominent physicist, Homi Bhabha, as its director.[36] In 1948, Jawaharlal Nehru, India's first Prime Minister, made an impassioned speech to the General Assembly of India advocating nuclear energy; later that year, an advisory board (the Atomic Energy Commission) was established under the Indian Ministry of Natural Resources and Scientific Research to further study the issue.[37] By August 1956, the first research reactor was operational, despite the accidental death of Bhabha.

The fledgling nuclear energy program was seamlessly connected to a vision of a prosperous and technologically advanced Indian society. Upon attaining independence, the Indian economy was dominated by the agrarian sector while the industrial sector was in a primitive state. From the outset, planners conceived of the national nuclear program as key to confirming the country's standing in the modern era, thus intersecting with the widely held belief that energy abundance underpinned social progress. Nehru argued in 1948 that India had failed to capitalize on the first Industrial Revolution due to lack of technical skill, and believed that success in the ongoing second Industrial Revolution was predicated on engineering prowess, typified by nuclear power. Later in the 1970s, Prime Minister Indira Gandhi reiterated Nehru's position that nuclear power was an essential technology for rescuing developing economies such as India's from "poverty and ignorance." She was convinced that a bold display of scientific and technological might could impress the populace enough to win her re-election.[38]

Japan, 1955–1990

Following defeat in World War II, much like France, Japan was in ruins. More than 30% of the Japanese population was homeless, communication and transport networks were in shambles, and industrial capacity had been bombed into insignificance.[39] With the support of Occupation funding, Japan embarked on a modernization program that would achieve unprecedented economic success. The promise of generating cheap energy through applied nuclear technology meshed perfectly with government aspirations to enhance the international competitiveness of industry. Japan's nuclear power program was officially launched when the government passed the Atomic Energy Basic Law in 1955, which set out the criteria under which peaceful development of nuclear technology was to be undertaken. Government development funding, which commenced that year, led to the inauguration of Japan's first nuclear energy plant, the Tokai Nuclear Power Plant, in 1966.

Japan's nuclear energy program was an offspring of aspirations for enhanced national energy security. National planners came to see nuclear technology as an important export product — a tool to not only free the nation from energy dependence, but also extend its economic reach into the Pacific and the world at large. The sheer lack of indigenous energy resources justified a massive expansion of the nuclear program, including commitment to plutonium-fueled fast breeder reactors. Japanese officials believed that a greater national risk was posed by dependence on imported energy than by a network of nuclear power plants.

Soviet Union, 1954–1986

The former Soviet Union was home to the first nuclear power plant in the world, a 5-MW graphite-moderated reactor at Obninsk that was built in 1954 and similar to the later design which failed at Chernobyl in 1986. Atomic energy was linked to visions of a radiant communist future. The one-party Communist system, its control over the media, and the suppression of doubts about science and technology provided an ideal environment for nuclear expansion.

Nuclear energy was quickly attached to the infallibility of Soviet science and technology, as well as the idea of a progressive communist regime free from energy shortages and wants. As a central slogan of the Soviet nuclear industry put it, "Let the atom be a worker, not a soldier."[40] Atomic energy came to represent not only a source of electricity supply for government planners, but also a pathway towards developing breeder reactors that would meet all of the country's energy needs, a first step towards perfecting nuclear-powered engines for aircraft and automobiles, a system for producing radiation to preserve food, a source of knowledge about nuclear technology that could help the Soviet Union build advanced weapons, and a mechanism of political control whereby planners dispersed nuclear reactors to the republics to strengthen ties and political adherence.[41] It also went hand-in-hand with an agenda to convert an agrarian and peasant society into a "well oiled machine of workers" tirelessly committed to communism.[42]

Early successes in nuclear research were seen as positive proof of the legitimacy of the entire way of Soviet thinking, and the promise of nuclear energy also reassured Soviet leaders about the concentration of the empire's energy reserves in Siberia and the Caspian Sea. Soviet engineers quickly became caught up in the fantasy of a nuclear Soviet Union, and spoke publicly about the applications of gamma ray mineral prospecting and oil surveying, the use of radiation for industrial monitoring and quality control, the creation of atomic fertilizers and viruses, and the irradiation of food and other items to prolong their shelf life. Soviet nuclear energy was "the instruction of nature at its finest"; and it was believed that widespread use would produce the energy needed to fill deserts with water, build canals, excavate waste sites, and accelerate industry. One plan even called for the melting and diversion of Siberian rivers so that the heavily populated Ukraine and Volga Basin regions could be irrigated.[43]

Nuclear power in the Soviet Union therefore fused together faith in Soviet science and technology, secrecy, defense, and gigantism.[44] Russian planners were captivated by science and technology, and became fascinated with the technology on display. Khrushchev encouraged Soviet scientists to "accelerate the construction of communism" by imitating Western methods of scientific experiment and management, culminating in the belief that atomic energy was almost a magical sort of alchemy. Radioisotopes were believed to help grow food quicker and cure diseases. This reaffirmed political control to an inner elite of party members, and created pressure for scientists to avoid delays in nuclear projects that could result in their arrest, dismissal, imprisonment, or even death. Nuclear energy was also pursued on security grounds to ensure parity with Western military might and secure Russian borders from invasion or interference; Soviet military planners spent billions of dollars researching nuclear-powered rockets, jets, ships, and satellites.

United States, 1942–1979

While the Soviet Union exhibited grand visions for nuclear energy, perhaps they paled in comparison to those in the US, where the atomic age began in December 1942 with an experiment at the University of Chicago and culminated in the completion of the Manhattan Project. By the end of World War II, planners were looking for civilian applications of the atom, and its possibilities were seen as endless. Scarcely one year after the War ended, Congress established the Atomic Energy Commission (AEC), which believed that atomic energy should not only enhance defense but also "promote world peace, improve the public welfare, and strengthen free competition in private enterprise."[45] The AEC was established as an executive agency with complete control over nuclear development and exclusive ownership of fissionable materials and all facilities. The creation of the AEC gave the federal government control and authority over all aspects of the technology. Put another way, the AEC was given "monopoly like powers protected by the cover of national security."[46] (This emphasis on peace is a bit ironic, given that, when the US Air Force discovered that the Soviet Union had detonated a nuclear device in September 1949, the civilian reactor program was intertwined with military efforts; generals hoped that civilian reactors could produce a "quantum jump" to develop a thermonuclear weapon.[47])

As one example of the hype surrounding nuclear energy, the same month the atomic bombs were dropped on Hiroshima and Nagasaki, the pocket book *The Atomic Age Opens* was published and widely read. The book depicted a future world in which coal and petroleum would go unused, and existing hydroelectric facilities would be abandoned and as "obsolete as the stagecoach" was in 1945. To give the general public some feeling for the vast amounts of energy soon to be theirs, the authors calculated the atomic power of ordinary things: one pound of water had enough energy to heat 100 million tons of water, a handful of snow could power an entire city, and the energy in a small paper railway ticket was sufficient to power a heavy passenger train several times around the earth.[48] Robert M. Hutchins, President of the University of Chicago, stated in 1946 that nuclear power would make "heat so plentiful that it will even be used to melt snow as it falls." Hutchins went on to suggest that "a very few individuals working a few hours a day at very easy tasks in the central atomic power plant will provide all the heat, light, and power required by the community and these utilities will be so cheap that their cost can hardly be reckoned."[49]

Nuclear energy promotion also reinforced national values and ideas about technology and nature. The anthropologist Gary Downey argues that advanced technology has always been correlated with progress in the US, and was initially used to distinguish the American colonies from their English counterparts. Thus, nuclear energy was seen as politically necessary to avoid the risks of communism, and was key to a postwar identity shaped in defiance to Marxism and Communism. Military planners believed that demonstrating the civilian applications of the atom would also affirm the American system of private enterprise, showcase the expertise of scientists, increase personal living standards, and defend the democratic lifestyle against Communist intrusion.[50]

Less than ten years after Hutchins' statement, the US government fully embraced nuclear power and passed the Atomic Energy Act of 1954 — the same year that President Dwight Eisenhower pledged to "strip the atom's military casing and adapt it to the art of peace."[51] The central theme behind the "Atoms for Peace" project was to show that the power of the atom could be converted from a terrifying military force to a benign commodity. The role of the government was to be a custodian of atoms.[52] Lewis Strauss, Chairperson of the AEC, remarked that atomic power would usher in an age where:

> It is not too much to expect that our children will enjoy in their homes electrical energy too cheap to meter, will know of great periodic regional famines in the world only as matters of history, will travel effortlessly over the seas and under them and through the air with a minimum of danger and at great speeds, and will experience a lifespan far longer than ours as disease yields and man comes to understand what causes him to age.[53]

Partially captivated by such optimism, Eisenhower's "Atoms for Peace" program granted US$475 million in funds to promote nuclear power abroad and Walt Disney even produced a television show entitled "Our Friend, the Atom."

One of the drivers behind atomic energy in the US was competition with the Soviet Union. Developments outside the nuclear industry during the 1940s and 1950s — such as the Alger Hiss case, the pro-Soviet coup in Czechoslovakia, the Soviet blockade of West Germany, the Chinese Revolution, as well as Soviet progress in developing atom bombs, hydrogen bombs, and nuclear reactors — convinced many American planners that they were in a "race to save the world from communism." Nuclear power was one key component of winning this race. It is illustrative that the first nuclear plant built by the AEC in Shippingport, Pennsylvania, started in 1953 directly after the Soviet Union exploded its H-bomb, and that the reason for choosing to go forward was not to produce a "cost-competitive" plant but to show the world that the US could design and operate a reactor.[54]

Spain, 1951–1980

Spain pursued a path of nuclear power partly because of its technocratic government, imperialist ambitions, utopian thinking, and Cold War relationships. Its quest for nuclear energy began in the early 1940s. After the atom bombs were dropped on Japan, Spanish leaders were convinced that military might lay in nuclear weapons, not in soldiers or ships. The country also happened to be sitting on what was believed to be one-seventh of the world's recoverable uranium deposits. Planners there established the *Junta de Energía Nuclear* (Nuclear Energy Board, or JEN) in 1951, and promoted nuclear power on the grounds that Spain had to be involved with important developments in science. As a consequence of its dictatorship and its collaboration with the Third Reich during World War II, Spain was excluded from international forums until 1955 and did not receive economic aid under the Marshall Plan. Impoverished by war, Spanish planners therefore saw nuclear energy as an inexhaustible source of energy necessary to power Spain's national reconstruction, development, and industrialization.[55]

Canada, 1942–1994

Canada's nuclear power industry can be traced back to uranium mining, which was initially under private control during World War II and operated to meet the needs of British and US military research. Under the 1943 Quebec Agreement, Canada funneled high-quality uranium to the Manhattan Project and clandestine British weapons programs; but when the war ended, the government declared all "works, undertakings, and substances relating to atomic energy to be for the general advantage of Canada." One year later, in 1944, construction began on an experimental research reactor. Canada later passed the Atomic Energy Control Act of 1946, which gave the government complete control over nuclear energy, expropriated all private uranium companies, established a Crown corporation (Eldorado Mining and Refining Limited), and prohibited all other actors from selling uranium in Canada to anyone other than this entity until 1959. Also, in 1952, Atomic Energy of Canada Limited was established as a government agency to coordinate research and regulate the export of nuclear materials and equipment.[56] The belief at the time was that Canada would be well positioned to supply the world fleet of reactors with uranium, making the country a de facto power broker in the transition to a global atomic economy.

Conclusion

In each of the above historical cases, planners pursued nuclear power not solely based on its costs and benefits in the present, but with hope about potential future gains, national visions, and technological optimism. As Table 4 shows, these visions differed by country and over time. Yet despite such differences, each of them painted nuclear energy as leading to national "radiance," economic revitalization, progress, and the possibility of a better future of some type. Their prevalence reminds us that energy policymaking is not always guided by coldly rational thinking alone, and that energy systems can play a forceful role in shaping norms and ideals about what the future may hold. However, it also illustrates that nuclear power was never initially designed or intended to be a cost-competitive source of electricity supply.

In the end, the choice between nuclear power and its cleaner alternatives boils down to a simple question: Do we want a nuclear economy, which is centrally administered by technical specialists, completely reliant on government subsidies, dependent on future breakthroughs in research, and sure to promote international proliferation and worsen inequity

Table 4:   **Visions Associated with Nuclear Power During the Formative Years of Eight National Programs**

Country	Period	National Vision
China	1953–1992	"Catching up" with other industrialized countries (including Taiwan), and creating lucrative opportunities for Chinese exports and economic leadership
France	1945–1970	Recovering from World War II and revitalizing the national economy through high-technology "national champions" that would legitimate France as a vital superpower
India	1945–1980	Creating a prosperous and technologically sophisticated Indian society in which social problems (such as hunger and poverty) would be eliminated
Japan	1955–1990	Using technological prowess and nuclear energy to rebuild the national economy, and to offset the risks of energy shortages and dependence on energy imports
Soviet Union	1954–1986	Validating the Communist system and the Soviet approach to science, and achieving a utopian future without scarcities of water, food, heat, or energy
United States	1942–1979	Harnessing the power of the atom for peaceful purposes, legitimizing the Manhattan Project, and creating a future in which electricity would be "too cheap to meter"
Spain	1951–1980	Revitalizing the Spanish economy after World War II and participating in "important" scientific research involving fission
Canada	1942–1994	Cultivating global demand for Canadian uranium and creating a lucrative export market for Canadian reactors

and vulnerability, that requires draconian security measures, wastefully generates and distributes electricity, remains based on highly uncertain projections about theoretical nuclear designs and available fuel, fouls water and the land, and trashes the planet for many future generations? Or, do we want a small-to-medium-scale decentralized electricity system, which is more efficient, independent from government funding, and encompassing commercially available technologies, that operates with minimal harm to the environment, remains resilient to disruptions and terrorist assaults, is equally available to all future generations, and is highly beneficial to all income groups?

When the true costs of nuclear energy are compared to the true benefits of renewable technologies, the answer is almost too obvious. In a carbon-constrained world, continued investment in nuclear technologies still on the drawing board makes little sense, especially as such technologies rely on diminishing stocks of usable uranium that will require more and more energy inputs in order to be enriched to fuel-grade status. Why invest in nuclear energy as a solution to global climate change when, by the time such systems come online, enriching the fuel for them will require emitting as much carbon as today's fossil fuel systems?

Any rational investor, regulator, and citizen would choose instead to invest in the deployment of technologies that require little to no energy inputs so as to harness free and clean fuels widely throughout the world. Policymakers should peek beyond the smoke-and-mirrors Kabuki dance used to obscure the obvious advantages of renewable technologies and the obvious costs of nuclear systems. Any effective response to electricity demand in a world facing climate change involves enormous expansion in our use of renewable technologies and a steady abandonment of nuclear power.

Endnotes

[1] E.S. Beckjord *et al.*, *The Future of Nuclear Power: An Interdisciplinary MIT Study* (Cambridge, MA: MIT, 2003).

[2] Vaclav Smil, "Energy in the Twentieth Century: Resources, Conversions, Costs, Uses, and Consequences," *Annual Review of Energy and Environment* 25 (2000), p. 46.

[3] Trevor Findlay, *The Future of Nuclear Energy to 2030 and Its Implications for Safety, Security, and Nonproliferation* (Waterloo, Ontario: Centre for International Governance Innovation, 2010), pp. 2 and 8.

[4] Quoted in Travis Madsen, Johanna Neumann, and Emily Rusch, *The High Cost of Nuclear Power: Why America Should Choose a Clean Energy Future over New Nuclear Reactors* (Baltimore: Maryland PIRG Foundation, March 2009), p. 9.

[5] R. Andreas Kraemer, "Presentation to the Transatlantic Agenda for Global Nuclear Governance," Potsdam, Germany, March 5, 2010, p. 2.

[6] A.B. Lovins, "Nuclear Power: Economics and Climate Protection Potential" (Snowmass, CO: Rocky Mountain Institute, 2005), available at http://www.rmi.org/images/other/Energy/E05-08_NukePwrEcon.pdf/ (accessed October 4, 2005).

[7] Richard W. Clapp, "Nuclear Power and Public Health," *Environmental Health Perspectives* 113(11) (2005), pp. 720–721.

[8] Quoted in Shahla M. Werner, "Nuclear Energy Too Risky When Efficiency Works," *Milwaukee Journal Sentinel*, May 9, 2009.

[9] Amory B. Lovins, Imran Sheikh, and Alex Markevich, "Forget Nuclear," *Rocky Mountain Institute Solutions* 24(1) (Spring, 2008), p. 27.

[10] Joseph Romm, *The Self-Limiting Future of Nuclear Power* (Washington, D.C.: Center for American Progress Action Fund, June 2008).

[11] John Carlin, *Environmental Externalities in Electric Power Markets: Acid Rain, Urban Ozone, and Climate Change* (Washington, D.C.: NARUC, 1993).

[12] Russell Lee, "Externalities and Electric Power: An Integrated Assessment Approach" (Oak Ridge, TN: Oak Ridge National Laboratory, 1995, CONF-9507-206—2).

[13] US Department of Energy and the Commission of the European Communities, "U.S.–EC Fuel Cycle Study: Background Document to the Approach and Issues," *Report No. 1 on the External Costs and Benefits of Fuel Cycles* (Oak Ridge, TN: Oak Ridge National Laboratory, November 1992, ORNL/M-2500).

[14] Thomas Sundqvist and Patrik Soderholm, "Valuing the Environmental Impacts of Electricity Generation: A Critical Survey," *Journal of Energy Literature* 8(2) (2002), pp. 1–18; and Thomas Sundqvist, "What Causes the Disparity of Electricity Externality Estimates?," *Energy Policy* 32 (2004), pp. 1753–1766.

[15] Jim Giles, "When the Price Is Right: Chernobyl and the Future," *Nature* 440 (2006), p. 984.

[16] Mycle Schneider, Steve Thomas, Antony Froggatt, and Doug Koplow, *The World Nuclear Industry Status Report 2009* (Paris: German Federal Ministry of Environment, Nature Conservation and Reactor Safety, August 2009, UM0901290).

[17] John Byrne and Steven M. Hoffman, "The Ideology of Progress and the Globalisation of Nuclear Power," in John Byrne and Steven M. Hoffman (eds.), *Governing the Atom: The Politics of Risk* (London: Transaction Publishers, 1996), pp. 17–18.

[18] Amory B. Lovins, "Energy Myth Nine — Energy Efficiency Improvements Have Already Reached Their Potential," in Benjamin K. Sovacool and Marilyn A. Brown (eds.), *Energy and American Society — Thirteen Myths* (New York: Springer, 2007), pp. 259–260; and Dan Watkiss, "The Middle Ages of Our

Energy Policy — Will the Renaissance Be Nuclear?," *Electric Light & Power* (May/June, 2008), pp. 12–18.

[19] Benjamin K. Sovacool and Christopher Cooper, "Nuclear Nonsense: Why Nuclear Power Is No Answer to Climate Change and the World's Post-Kyoto Energy Challenges," *William & Mary Environmental Law and Policy Review* 33(1) (2008), pp. 1–119.

[20] Madsen *et al.* (2009).

[21] Aarti Dhar and J. Venkatesan, "Limiting Nuclear Liability Is a Violation of Rights: Sorabjee," *The Hindu* (December 11, 2009), p. 12.

[22] Marshall Goldberg, *Federal Energy Subsidies: Not All Technologies Are Created Equal* (Washington, D.C.: Renewable Energy Policy Project, July 2000, Report No. 11).

[23] Sovacool and Cooper (2008).

[24] Matthew L. Wald, "In Finland, Nuclear Renaissance Runs into Trouble," *New York Times,* May 29, 2009.

[25] Steven Mufson, "Nuclear Projects Face Financial Obstacles," *Washington Post,* March 2, 2010, p. A1.

[26] Madsen *et al.* (2009).

[27] Judy Pasternak, "Nuclear Energy Lobby Working Hard to Win Support," *McClatchy Newspapers,* January 24, 2010.

[28] Otis Dudley Duncan, "Sociologists Should Reconsider Nuclear Energy," *Social Forces* 57(1) (September, 1978), pp. 1–22.

[29] Byrne and Hoffman (1996), pp. 11–46.

[30] Martin Melosi, "Energy Transitions in Historical Perspective," in Laura Nader (ed.), *The Energy Reader* (London: Wiley-Blackwell, 2010), pp. 45–60.

[31] Michael G. Gallagher, "Nuclear Power and Mainland China's Energy Future," *Issues and Studies* 26(12) (1990), pp. 100–120.

[32] For an excellent history of nuclear power in France, see Gabrielle Hecht, *The Radiance of France: Nuclear Power and National Identity After World War II* (Cambridge, MA: MIT Press, 1998); and L. Scheinman, *Atomic Energy Policy in France Under the Fourth Republic* (Princeton: Princeton University Press, 1965).

[33] Wolfgang Rudig, "Outcomes of Nuclear Technology Policy: Do Varying Political Styles Make a Difference?," *Journal of Public Policy* 7(4) (1988), pp. 389–430.

[34] See Gene I. Rochlin, "Broken Plowshare: System Failure and the Nuclear Power Industry," in Jane Summerton (ed.), *Changing Large Technical Systems* (San Francisco: Westview Press, 1994), pp. 231–261; and Michael T. Hatch,

"Nuclear Power and Postindustrial Politics in the West," in John Byrne and Steven M. Hoffman (eds.), *Governing the Atom: The Politics of Risk* (London: Transaction Publishers, 1996), pp. 201–246.

[35] Hecht (1998), p. 2.

[36] David Hart, *Nuclear Power in India: A Comparative Analysis* (London: George Allen & Unwin, 1983).

[37] See *ibid.*; and Manu V. Mathai, "Elements of an Alternative to Nuclear Power as a Response to the Energy-Environment Crisis in India," *Bulletin of Science, Technology, & Society* 29(2) (April, 2009), pp. 139–150.

[38] Byrne and Hoffman (1996), pp. 11–46.

[39] J.W. Hall, *Japan: From Prehistory to Modern Times* (Tokyo: Charles E. Tuttle Publishers, 1990).

[40] Paul R. Josephson, *Red Atom: Russia's Nuclear Power Program from Stalin to Today* (New York: W.H. Freeman, 1999), p. 38.

[41] Josephson (1999).

[42] Paul R. Josephson, "'Projects of the Century' in Soviet History: Large-Scale Technologies from Lenin to Gorbachev," *Technology and Culture* 36(3) (July, 1995), pp. 519–559.

[43] *Ibid.*

[44] Josephson (1999); and *ibid.*

[45] Alice L. Buck, *A History of the Atomic Energy Commission* (Washington, D.C.: US Department of Energy, July 1983, DOE/ES-0003/1), p. 1.

[46] Lee Clarke, "The Origins of Nuclear Power: A Case of Institutional Conflict," *Social Problems* 32(5) (June, 1985), p. 477.

[47] See Buck (1983); and *ibid.*, pp. 474–487.

[48] Editors of Pocket Books, *The Atomic Age Opens* (New York: Pocket Books, 1945), pp. 202–203.

[49] Daniel Ford, *Meltdown: The Secret Papers of the Atomic Energy Commission* (New York: Simon & Schuster, 1986), p. 30.

[50] Gary L. Downey, "Risk in Culture: The American Conflict over Nuclear Power," *Cultural Anthropology* 1(4) (1986), pp. 388–412.

[51] Richard Munson, *From Edison to Enron: The Business of Power and What It Means for the Future of Electricity* (London: Praeger, 2005), p. 80.

[52] Shelia Jasanoff and Sang-Hyun Kim, "Containing the Atom: Sociotechnical Imaginaries and Nuclear Power in the United States and South Korea," *Minerva* 47(2) (2009), pp. 119–146.

[53] Lewis Strauss, "Speech to the National Association of Science Writers, September 16th, 1954," *New York Times* (September 17, 1954), p. 1A.

[54] Clarke (1985), pp. 474–487.

[55] Albert Presas I. Puig, "Science on the Periphery: The Spanish Reception of Nuclear Energy," *Minerva* 43 (2005), pp. 197–218.

[56] Constance D. Hunt, "Canadian Policy and the Export of Nuclear Energy," *University of Toronto Law Journal* 27 (Winter, 1977), pp. 69–104.

Postscript: The "Hydra-Headed" Fukushima Nuclear Crisis

In ancient Greek mythology, the hydra was a serpent-like beast with many heads that guarded the entrance to the Underworld. Its breath was so deadly that even its footprints were reputed to poison men to death. In the past few weeks, the nuclear industry may have met its hydra in the form of an accident in the Fukushima Prefecture of Japan. Chapter 3 of this book on "normal accidents" documented historical safety and reliability issues with nuclear power plants, and also predicted that at least four serious core damage accidents would occur between 2005 and 2055. One of these four accidents has just happened at the Fukushima Daiichi nuclear power plant in Japan, where an earthquake and tsunami have caused emergency backup generators to fail and the pressure vessels at some of its reactors to explode. Spent fuel pools at the facility have caught fire, fuel assemblies have melted down, and dangerous levels of radiation have been reported. At the time of writing, more than 200,000 residents have been evacuated from a 30-km safety zone and 160 people have been exposed to "hazardous" levels of radiation, in addition to 21 fatalities (7 from first responders and plant operators and 14 elderly persons killed during the evacuation).

The Fukushima Daiichi facility houses six of Japan's 55 reactors and is one of Japan's 17 nuclear power plants, which together generate about

30% of the country's electricity. The Fukushima Daiichi plant, 150 miles (240 km) outside of Tokyo, relies on boiling water reactors which circulate water through the core and convert it into steam to drive an electric generator. The first unit at Daiichi was connected to the grid in November 1970 and the sixth unit was connected in March 1979. Owned and operated by the Tokyo Electric Power Company (Tepco), each reactor is loaded with 12-feet-long fuel rods made of radioactive uranium-235 pellets surrounded by zirconium alloy, with some relying on more toxic mixed oxide (MOX) fuel containing a mixture of uranium-235 and plutonium.

Chronology of Events

The story so far — and it is still unfolding — starts with a 9.0-magnitude earthquake that struck 370 km northeast of Tokyo, 24.5 km below the ocean, at 2 a.m. on Friday, March 11, 2011.[1] The quake, the fifth largest in the world on record, sent 12-meter-high tsunami waves sweeping across rice fields, tossing cars and boats, engulfing entire towns, and reaching as far as 10 km inland in some places such as Miyagi Prefecture. The quake also induced substantial damage at Tepco's Fukushima Daiichi power plant, forcing plant operators to begin an emergency shutdown.

This shutdown involved moving control rods below three of the six operating reactors — the other three had been closed for maintenance at the time — into the core to absorb neutrons and stop the chain reaction. The shutdown proceeded as planned within minutes of the quake, but other reactions in the reactor still produced substantial decay heat. When the tsunami waves crashed into the power plant one hour later, the shutdown reactors were still as powerful as a commercial jet engine at full throttle in a very confined space.

The tsunami itself physically washed away all of the plant's backup diesel generators, and its batteries provided enough electricity to cool the facility for only a few hours. It was then that all three reactors began to melt. Fukushima Daiichi's boiling water reactors are not unlike an electric kettle that operates at 550°F, below the temperature of a coal furnace and slightly hotter than an ordinary kitchen oven.[2] The nuclear reactor is like the part at the bottom of the kettle that heats the water. If the kettle cannot be turned off, and the spout is sealed because the steam is radioactive, then the amount of water around the reactor will slowly decrease, exposing the reactor core. In the case of Fukushima, all three reactor cores started to melt to the bottom of the steel pressure vessel and react with the steam, causing temperatures to rise above 5,000°F.[3] These high temperatures compromised the reactor vessel and surrounding containment systems, releasing radioactive materials throughout the plant.

On Saturday, March 12, operators at the facility responded to the melting reactors by venting some of this radioactive steam to reduce the stress on containment structures. However, the resulting hydrogen gas somehow found a spark and exploded, blowing the roofs off the buildings surrounding both reactors 1 and 3 and severely injuring four workers. Because the cooling systems for each reactor had malfunctioned, the operators began to flood them with seawater laced with boric acid.

On Sunday, March 13, authorities started detecting abnormally high radiation levels around the plant and in the nearby prefecture, and distributed iodine tablets to residents. Three people randomly selected from a group of 90 tested positive for radiation exposure, and Chief Cabinet Secretary Yukio Edano cautioned that other explosions might be eminent.

On Monday, March 14, things became worse. Multiple explosions, again believed to come from hydrogen gas, occurred at reactors 2 and 3, further damaging what was left of the multilayered cooling systems and containment vessels, injuring 14 other workers, and exposing 190 workers to unsafe levels of radiation. About 80,000 residents within a 20-km radius of the plant were evacuated, and up to 2.8 meters of the control rods in reactor 2 were left uncovered because the pump keeping them cool failed. The reactor began to emit radioactive steam. Radiation levels at the plant exceeded 11,000 μSv per hour, enough to cause radiation sickness (400 μSv can cause temporary sterility in men). Later that evening, reactor 2 boiled completely dry, causing another explosion at the bottom of the containment vessel, leading to the fire and steam shown in Figure 1.

On Tuesday, March 15, another explosive impact shook reactor 2, damaging its suppression pool. The US Navy began repositioning its ships used in humanitarian disaster relief away from "airborne radioactivity" it had detected in the region. (Three people aboard the *USS Ronald Reagan* tested positive for low levels of radiation.) The Japanese government extended the evacuation zone to 30 km and asked roughly 180,000 residents

Figure 1:   Radioactive Steam from Fukushima Daiichi's Reactors 2 and 3

to leave the area. Tepco also evacuated almost all of the plant's 800 staff, leaving only 50 workers to handle the crisis. Later that morning, a fire broke out at the cooling pond used for nuclear fuel at reactor 4, which had been shut down, venting radioactive iodine and cesium directly into the environment. Tepco prepared an emergency plan to pour water dumped from military helicopters onto the reactors. The closely spaced explosions released a surge of radiation within the plant 800 times as intense as the recommended exposure limit.

On Wednesday, March 16, a second fire broke out at reactor 4, dispersing more radioactive material into the atmosphere. Radiation levels soared inside and around the plant, and also began to rise as far as 200 km away. Indeed, radiation levels were reported to be 20 times higher than normal in Tokyo.

Since then, no more fires and explosions have been reported, although radioactive steam has emanated from the facility for the past two weeks. As of today, March 28, the government has set up a 30-km "no-fly zone" and containment perimeter as shown in Figure 2, and Prime Minister Naoto

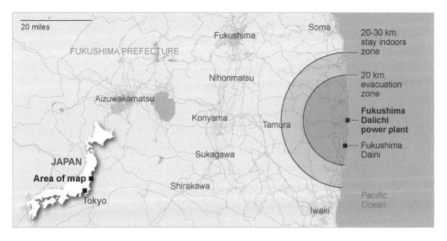

Figure 2: The 30-km Exclusionary Zone Around the Fukushima Daiichi Nuclear Power Plant[5]

Kan stated over the weekend that the situation at the Daiichi facility was "very grave and serious," with Tepco officials reporting that the containment vessel at reactor 3 had been breached, leading to "severe radioactive contamination."[4] Radiation levels at the spent fuel ponds that caught fire have been reported to be 10 million times above the normal limit due to radioactive iodine-134; and surface water nearby has shown 1,000 mSv of radiation, far above the safe limit of 3 mSv. A monitoring post nearby has also measured radiation levels in the sea 1,850 times higher than normal. Traces of radioactive iodine have been detected as far away as Heilongjiang Province in China.

Broader Implications

In the wake of Fukushima, Germany has already declared a three-month moratorium on its plan to prolong the life of its nuclear plants and has shut down seven of its oldest facilities. Switzerland has announced that it is reassessing its nuclear program and has suspended plans to replace its nuclear reactors. Chinese planners have also stopped approval for all nuclear power plants, and have halted all plants in construction (though it is unclear how long this moratorium will last).[6] While it is difficult to measure the broader implications of the accident so close to its occurrence, at least five preliminary conclusions can be drawn.

The Accident Is the Worst Since Chernobyl in 1986

The Fukushima accident is worse than the Three Mile Island accident in 1979, but not yet as bad as the Chernobyl accident in 1986. Operators at Three Mile Island (TMI) were able to avoid a hydrogen explosion; whereas multiple explosions and fires have occurred at multiple parts of the Daiichi complex, from its reactor vessels to its spent fuel ponds. At TMI, only modest levels of radiation were reported and no plant operators immediately died, yet at Fukushima seven operators have already died. At TMI, only one reactor was in trouble; whereas four of Fukushima's reactors have suffered melting fuel, loss of coolant, and/or exposed fuel cores. Currently, the International Nuclear and Radiological Event Scale classifies the accident as level 6, one step down from the most serious level 7 (Chernobyl) and above level 5 (TMI).

The Accident Has Revealed a Culture of Secrecy, Cover-Ups, and Misinformation

Immediately after the earthquake, Tepco reported that efforts to shut down the reactors had been successful and that there was nothing to worry about. Even while the reactor cooling systems were failing and radiation levels were spiking around the plant, Yukio Edano, the chief government spokesman, stated that "there are no reports of leakage from any nuclear-power plants at the moment and no signs of any leakage."[7] Similarly, on Saturday, March 12, when hydrogen explosions destroyed containment structures, authorities insisted that no harmful gases had been released and that the explosion had released only "water vapor that was part of the cooling process," despite the fact that monitoring stations around the plant showed that radioactive cesium and iodine had escaped containment structures. Even the Prime Minister apparently was not getting the full story from Tepco, and somewhat infamously marched down to their offices to demand to know "what the hell is going on."[8] Tepco officials did not release knowledge about radioactive exposure until the United States Seventh Fleet publicly announced that the

USS Ronald Reagan and *USS George Washington*, more than 280 km away, had encountered a low-level radioactive plume. The US Nuclear Regulatory Commission accused Tepco of trying to downplay the seriousness of the accident and setting an insufficient safety perimeter.[9] Tepco and the Japanese Ministry of Health, Labor, and Welfare also raised the maximum limit on exposure for workers in an emergency to 250,000 µSv, 2.5 times above the previous limit, so they could ingeniously claim that workers were still operating within "government limits."[10]

Already, past incidents involving Tepco and the plant itself have come to surface. In 2002, Tepco's vice president and chairman resigned after a scandal in which the utility was accused of falsifying safety repair records in 29 cases. In 2003, Tepco had to shut down all 17 of its nuclear power facilities after it emerged that they had covered up reports showing cracks in the structures of some reactors.[11] Tepco was also repeatedly warned that their power plants were not built to withstand earthquakes greater than 6.5 on the Richter scale, and needed special equipment. In July 2007, another earthquake shut down seven reactors, three of them permanently, at Tepco's Kashiwazaki plant on the west coast of Japan. Tepco was yet again urged to upgrade its safety plans. In December 2008, the International Atomic Energy Agency (IAEA) specifically warned that seismic safeguards at Japanese nuclear power plants were "outdated and inadequate."[12]

These defects become especially pertinent when one realizes that the Fukushima Daiichi plant was not the only facility damaged in the quake. Initially after the tsunami, five other nuclear power plants in Japan declared a "state of emergency." Cooling systems malfunctioned at three reactors at the Fukushima Daini facility about 80 km south of Sendai and close to the Daiichi facility, forcing operators to vent radioactive steam to reduce pressure and also to declare a 10-km evacuation zone. This facility was much newer than Daiichi, having been connected to the grid in 1981 and 1987. Excessive radiation levels and a fire in the turbine house were recorded at the Onagawa facility about 70 km north of Sendai, and its reactors entered service between 1983 and 2001 (thus partially refuting the idea that Daiichi's age made it vulnerable to the quake). Cooling pumps damaged by the earthquake also failed at the Tokai nuclear power plant in Ibaraki Prefecture.

At the Fukushima Daiichi plant itself, some of its reactors were supposed to be decommissioned because of design faults but had their operating licenses extended for 10 years at the request of Tepco.[13] Tragically, in 2002, an advisory group recommended that Tepco raise its maximum projected tsunami level at Fukushima Daiichi and increase the height of its backup generators, but the company merely responded by raising a single pump by eight inches.[14] In October 2010, the Nuclear Safety Commission of Japan cautioned Tepco that their facilities, including Daiichi, were at a "residual risk concerning earthquakes and tsunamis" and that cooling systems needed to be retrofitted.[15] The Commission also warned that backup diesel generators were vulnerable to corrosion from seawater and rainwater.[16] Moreover, workers at the Daiichi facility failed to inspect about three dozen pieces of key cooling system equipment ten days before the March 11 earthquake.[17] One Tepco engineer has even come forward to argue that the Daiichi reactors may have been relying on flawed steel in its containment vessel, and that he had warned plant officials of a "time bomb" but was ignored.[18]

As one commentator pointed out, "This is an industry with a long record of cover-ups of dangerously damaged facilities, and cover-ups of safety violations, and unreported radioactive leaks, and inadequate waste storage protections, and napping guards, and more radioactive leaks, and more radioactive leaks, and on and on."[19] *The Economist* also wrote that the "country's nuclear industry has a long history of cover-ups and incompetence"[20] and "a shameful record of cover-ups, lackadaisical crisis management, and an inbred complicity between regulators and utilities."[21] Worryingly, these problems are not unique to Japan — it emerged a week after the Fukushima accident that almost 30% of nuclear power plants in the US had concealed defects, bungled repairs, and failed to report safety violations, suggesting that the trend may be industry-wide.[22]

Human Error Played a Key Role in Exacerbating the Accident

Human errors in design, operations, maintenance, and emergency response certainly played their part in causing and worsening the situation. The plant was not designed to withstand a 9.0-level earthquake, and it was also constructed to house all six of its reactors in close proximity to each other in order to reduce costs and make moving equipment easier. However, the tsunami washed over Daiichi's backup generators, explosions at the first reactor hindered efforts to cool the other reactors, and multiple reactor fires distracted operators from cooling the spent fuel ponds. The Daiichi facility stores its spent fuel within the reactor building to make loading and unloading simpler, but this meant that meltdowns and fires affected both reactors and spent fuel simultaneously.[23] When the cooling systems failed at Daiichi, military helicopters attempted to dump water from the air and firefighters used water cannons; unfortunately, these did not cool the reactors as planned, since they were designed for forest fires and riot control.[24]

As one engineer admitted, "the earthquake and tsunami we had last week both exceeded our engineering assumptions by a long shot."[25] A former director of the Fukushima Daiichi plant stated that "we can only work on precedent, and there was no precedent. . . . [W]hen I headed the plant, the thought of a tsunami never crossed my mind."[26] Daiichi relied on reprocessed MOX fuel to minimize the need for fresh uranium ore; however, this meant that when reactor 3 failed, its plume was more dangerous because it had greater amounts of plutonium. When the fuel cladding started to melt at reactors 1 and 3, workers had to evacuate the control rooms as radiation levels were 1,000 times above the safe level, making it even more difficult to monitor and control events — another contingency that had not been anticipated.

Ultimately, it is the human element that adds a degree of unforeseen risk to any nuclear power accident. As Matthew Bunn from Harvard University has written:

> [With Fukushima Daiichi,] people have not adequately thought through the possibility of multiple traumas that could be caused by the same initiating event. . . . This reinforces the view that whenever someone says there is less than a one-in-a-million chance of a complex system failing, there is more than a one-in-a-million chance they have made unjustified assumptions in their estimate.[27]

In Japan, it was not the "nuclear parts" of the station that went wrong initially, but "conventional ones" such as pumps and backup generators. Or, as John Vidal has concluded, "it's easy to be wise afterwards, but the inquest will surely show the accident was not due to an unpredictable natural disaster, but by a series of highly predictable bad calls by human regulators."[28] The sobering lesson appears to be that it is impossible to design a nuclear power plant for all unknowns.

The Full Cost of the Accident Will Be Quite Large

We already know that the Fukushima Daiichi accident will bring with it a suite of terrible technical, economic, environmental, and sociopolitical costs.

The severity of damage at the power plant itself is quite significant: the moment operators started pumping seawater to cool the reactors, they became too corroded to ever generate electricity again. Tepco has estimated that it will need at least US$25 billion in loans for repairs and to find new sources of electricity to replace the Daiichi plant.[29] This is to say nothing of the lost revenue from the plant, the sunk costs in the facility, and the financial burden of having to contain and decommission it. This last phase will likely be as expensive as the multibillion-dollar operation at Chernobyl, where the reactor site was entombed in concrete and buried in sand.[30] Already at least 7 plant workers have died and 47 have been injured in the explosions and fires,[31] in addition to 14 elderly persons who died during the evacuation as they were moved from hospitals, and many more have suffered from severe radiation exposure.[32] Some workers have already been exposed to radiation levels high enough to cause sterility and hemorrhaging; and they have required treatment for radiation sickness, nausea, vomiting, diarrhea, and plummeting blood counts.[33] The most recent breach of the reactor on March 25 exposed workers to 100,000 times more radioactivity than normal.

More broadly, the accident has had economy-wide implications such as blackouts throughout Tokyo, the collapse of the Japanese stock market, and rising prices for coal and natural gas. Rotating power outages due to the closure of Fukushima Daiichi have occurred in Tokyo and eight other prefectures, with more than 10% of Japanese households without electricity. These blackouts have depressed economic output and made it harder for the economy to recover from the earthquake. Panic, especially over radiation, has hit the financial markets, with Japanese stocks falling by 6.2% on the day of the quake and by 12% on the day the partial meltdowns were revealed. Due to a combination of panic and lack of electricity, Japanese auto manufacturers have closed down factories, airlines have canceled their flights to Japan, and US$364 billion of wealth has been wiped off the Tokyo Stock Exchange. Tepco saw its shares drop by 25%, while Toshiba's fell by 20%.[34] Moreover, global prices for liquefied natural gas have risen as more cargoes are diverted from Europe to Japan to make up for its shortfall in electricity.[35] Global uranium prices have dropped by 25%, while gas and coal prices have increased by 13.4% and 10.8%, respectively, underscoring the global ramifications of Fukushima.[36]

Environmentally, the radioactive plume from Fukushima has spread far beyond Japan. Radiation levels in Tokyo were recorded to be 100 times higher than normal (measured at 5,575 μSv per hour), and the IAEA has measured potentially harmful levels of radiation as far as 130 miles away from the facility. Figure 3 shows a child being tested for radiation in Fukushima Prefecture. Radiation 1,600 times higher than normal has also been detected far off the coast of Japan, most likely from contaminated

Figure 3: A Family Being Tested for Radiation Exposure in Fukushima Prefecture, March 12, 2011

cooling water that has gradually drained off the Daiichi site. The city of
Iwaki has suffered radioactive rain and has told its 340,000 citizens to stay
indoors; city planners fear that radiation from Fukushima has contaminated its reservoirs of drinking water.[37] The Health Ministry of Japan has
reported excess amounts of radioactive elements on canola, chrysanthe-
mum greens, and spinach, as well as contaminated milk at 37 dairy farms
in four prefectures.[38] The Health Ministry has also warned that tainted
food has likely already been sold and consumed. Radioactive iodine-131
has been detected in Tokyo's water supply, forcing government spokesper-
sons to declare it unsafe for infants and to start distributing bottled water.[39]
Taiwan has detected radiation on imported fava beans from Japan; and
Singapore has detected radioactive substances on imported Japanese wild
parsley, rapeseed, mustard, and perilla leaf from four prefectures, including
two far away from the Fukushima plant.[40] More than three dozen countries
have banned Japanese imports of vegetables, fruit, dairy products, meat,
and fish.[41] Meanwhile, a low-level radioactive plume has already spread
across the Pacific Ocean, the Aleutian Islands in Alaska, and the West Coast
of the United States.[42] Scientists at the Comprehensive Nuclear-Test-Ban
Treaty Organization expect the plume to circulate the world in 10–15
days.[43] Residents in Russia have been buying potassium iodide pills, as have
hundreds of thousands of people residing in coastal Chinese cities.

A final social and political impact, although hard to measure, has been
a delayed humanitarian response to the earthquake, fear and anxiety, and
panic buying. Disaster relief and humanitarian efforts following the earthquake and tsunami have been hampered by worries over radiation and
uncertainty over the Fukushima accident. Many Fukushima residents
claim that they would have fled the affected nuclear area, but did not have
any fuel to make the journey. They have also reported that outsiders were
not willing to come and get them because of fears of radiation.[44] Relief
organizations such as the Red Cross Society and Tokyo Fire Department
have suspended and withdrawn some of their evacuation and emergency
response operations due to radiation concerns.[45] Fears of a complete meltdown have prompted panic buying of basic necessities in Tokyo and
elsewhere, with supermarket shelves empty and shortages of basic goods.
(To be sure, such fear and anxiety may have been worse in other cultures;
most people in Japan have been stoic and resilient in the face of

Fukushima, with little complaining among evacuees and Tokyo people still waiting in line to pay their taxes two days after the quake struck.) Some corporations and embassies have also asked their citizens to move from affected areas in Japan, with the US State Department approving the departure of government personnel and several European countries urging their nationals to leave Japan altogether.

The Accident Was Unnecessary Given Japan's Renewable Resource Base

Although the vision of hindsight is often perfect, the Fukushima accident was avoidable insofar as Japan could have chosen instead to invest entirely in renewable energy resources. In 2009, Japan had roughly 290 GW of installed electrical capacity. Many experts and Japanese policymakers have argued that nuclear power is "unavoidable" for the country, given its lack of domestic resources. Yet, Japan has a total of 324 GW of achievable potential in the form of onshore and offshore wind turbines (222 GW), geothermal power plants (70 GW), additional hydroelectric capacity (26.5 GW), solar energy (4.8 GW), and agricultural residue (1.1 GW).[46] If policymakers had embarked upon a path of investing in these renewables instead of nuclear facilities, the Fukushima Daiichi accident would never have occurred. Japan certainly has enough renewable energy potential to displace all 50 GW of its nuclear power plants, let alone every existing conventional power plant connected to the grid in 2010.

Conclusion

For nuclear power to be safe, for all risks to be known, Japan and other countries embracing nuclear energy need not just properly designed plants and safety procedures, but also good governance, accountability, and transparency. The Fukushima Daiichi nuclear accident was born from the opposite conditions, in a culture of secrecy, incompetence, and cover-ups. Perhaps one of the scariest things about the accident is that it is not a worst-case scenario; in Japan, far less than 1% of the radioactivity within the reactors was released and only 5% of its nuclear fuel was damaged. (Even for Chernobyl, only 3–4% of the radioactivity in the reactor core was released.) The grave and growing consequences of Fukushima — upwards of US$25 billion in damages to the plant, plus another few billion dollars for containment and decommissioning, 21 deaths and scores of injured and irradiated workers, rotating electricity blackouts, jittery global financial markets and rising fossil fuel prices, contaminated food and water, and compromised humanitarian relief for the earthquake and tsunami, to name a few — are *not* the worst that nuclear power plants can do when they malfunction. One does not need a lot of damage to a reactor or spent fuel pond to cause widespread misery. Taken together, the nuclear crisis, earthquake, and tsunami have been called a "triple disaster" that constitute what the Japanese are now calling the worst national crisis since World War II.

As *The Economist* eloquently summed up in their special briefing of the accident:

> Simply put, you can't trust the stuff. Somewhere, eventually, reactors will get out of control. One did at Three Mile Island in Pennsylvania in 1979. One did at Chernobyl in 1986. Now three have done so again . . . a bit like three Three Mile Islands in a row, with added damage in the spent-fuel stores. . . . Nuclear power thus looks dangerous, unpopular, expensive and risky. It is replaceable with relative ease and could be forgone with no huge structural shifts in the way the world works.[47]

The next nuclear disaster — and there will be one, as long as plants remain operating — may have nothing to do with an earthquake or tsunami, but may be caused by a terrorist attack, or a flood, or a design flaw, or a volcano, or simple human error. If there were no alternatives to nuclear power, then perhaps its collection of risks would be tolerable. But when so many viable alternatives exist that happen to be cheaper, less damaging, less dependent on subsidies, and safer, we do not need a world with more Fukushima meltdowns. Given the systemic risks involved with nuclear power, this book argued that a nuclear renaissance was unlikely to occur before Fukushima. Now, we have yet another strong reason to abandon nuclear power facilities in favor of energy efficiency and renewable energy technologies.

As one article proclaimed, the Fukushima crisis has been a problem of "hydra-headed complexity."[48] The analogy is apt, for in many ways the dangers of Fukushima are akin to the poisonous footprints of the hydra: invisible, implacable, and deadly. To successfully defeat the hydra in Greek legend, Heracles had to cut off all of its heads. Perhaps we will never be truly safe from accidents like Fukushima until we do the same to the nuclear industry.

Endnotes

[1] Most of this chronology is taken from Bryony Jones, "Timeline: How Japan's Nuclear Crisis Unfolded," CNN, March 15, 2011.

[2] William Tucker, "Japan Does Not Face Another Chernobyl," *Wall Street Journal*, March 15, 2011, p. 17.

[3] "Japan's Nuclear Industry: The Risks Exposed," *The Economist*, March 19, 2011, pp. 27–28.

[4] "Suspected Breach in Nuclear Plant Reactor," *Weekend Today*, March 26–27, 2011, p. 1.

[5] Source: Hiroko Tabuchi and Keith Bradsher, "Japan Says 2nd Reactor May Have Ruptured with Radioactive Release," *The New York Times*, March 16, 2011.

[6] Leslie Hook, "China Halts Nuclear Projects," *Financial Times*, March 17, 2011, p. 1.

[7] Yuka Hayashi and Andrew Morse, "Japan Officials Make Gains as Nuclear Crisis Sparks Rift," *Wall Street Journal*, March 16, 2011, p. 1.

[8] "Radiation Danger Rises," *The Straits Times*, March 16, 2011, p. 1.

[9] Norimitsu Onishi, David E. Sanger, and Matthew L. Wald, "U.S. Calls Radiation 'Extremely High'; Sees Japan Nuclear Crisis Worsening," *The New York Times*, March 16, 2011.

[10] Mitsuru Obe, "Workers Injured at Fukushima Daiichi Plant," *Wall Street Journal*, March 24, 2011.

[11] Norihiko Shirouzu and Alison Tudor, "Nuclear Troubles Put Spotlight on Japan Regulators," *Wall Street Journal*, March 15, 2011, p. 4.

[12] Lawrence Lewis, "It's Time to Leave Nuclear Power Behind," *Daily Kos*, March 20, 2011.

[13] Sofiah Jamil and Mely Caballero-Anthony, "Japan in Disaster: Managing Energy Vulnerabilities," March 15, 2011, available at http://www.rsis.edu.sg/publications/Perspective/RSIS0412011.pdf/.

14 Norimitsu Onishi and James Glanz, "Nuclear Rules in Japan Relied on Old Science," *The New York Times*, March 26, 2011.

15 Norihiko Shirouzu and Peter Landers, "Japan Ignored Warning of Nuclear Vulnerability," *Wall Street Journal*, March 23, 2011.

16 Hiroko Tabuchi, Norimitsu Onishi, and Ken Belson, "Japan Extended Reactor's Life, Despite Warning," *The New York Times*, March 21, 2011.

17 Associated Press, "Radiation Reaching Further into Japan's Food Chain," March 21, 2011, p. 3.

18 Jason Clenfield, "Fukushima Engineer Says He Helped Cover Up Flaw at Dai-Ichi Reactor No. 4," Bloomberg News, March 23, 2011.

19 Lewis (2011).

20 "The Fallout," *The Economist*, March 19, 2011, p. 13.

21 "Briefing Japan's Catastrophes: Nature Strikes Back," *The Economist*, March 19, 2011, pp. 24–26.

22 Tennille Tracy, "Nuclear Regulatory Commission: Plants Fail to Report Defects," *Wall Street Journal*, March 24, 2011.

23 Arjun Makhijani, "Post-Tsunami Situation at the Fukushima Daiichi Nuclear Power Plant in Japan: Facts, Analysis, and Some Potential Outcomes," March, 2011, available at http://www.ieer.org/comments/Daiichi-Fukushima-reactors_IEERstatement.pdf/.

24 Onishi *et al.* (2011).

25 Hiroko Tabuchi, David E. Sanger, and Keith Bradsher, "Nuclear Plant in Japan on the Brink," *International Herald Tribune*, March 16, 2011, p. 1.

26 Onishi and Glanz (2011).

27 Matthew Bunn, "Japan's Nuclear Power Plant Crisis," *The New York Times*, March 16, 2011.

28 John Vidal, "Japan's Avoidable Accidents Make Folly of Nuclear Energy Clear," *The Guardian*, March 15, 2011.

29 Atsuko Fukase, "Plant Operator Seeks $25 Billion," *Wall Street Journal*, March 24, 2011.

30 Reuters, "Nuclear Plant May Be Buried," *New Straits Times*, March 19, 2011, p. 4.

31 Keith Bradsher and Hiroko Tabuchi, "Last Defense at Troubled Reactors: 50 Japanese Workers," *The New York Times*, March 15, 2011.

32 Yuka Hayashi, "Japan Races to Dig Out, Cool Reactors," *Wall Street Journal*, March 15, 2011, pp. 1 and 7.

33 Yuka Hayashi, "Japan Faces Setbacks at Reactors," *Wall Street Journal*, March 17, 2011, p. 1.

34 Patrick Barta, Yoshio Takahashi, and Bob Davis, "Japan Woes Add to Worries Around Global Economy," *Wall Street Journal*, March 16, 2011, p. 1.

35 Guy Chazan, "Japan to Use More LNG," *Wall Street Journal*, March 16, 2011, p. 25.

36 Philip Stafford, Javier Blas, and Jack Farchy, "Nuclear Problems Put Energy Markets in a Spin," *Financial Times*, March 17, 2011, p. 15.

37 Phred Dvorak, "One City Isn't Taking Any Chances," *Wall Street Journal*, March 16, 2011, p. 4.

38 Associated Press (2011).

39 David Jolly and Kevin Drew, "Radiation Prompts Water Warning for Tokyo," *International Herald Tribune*, March 24, 2011, p. 1.

40 Leong Week Keat, "Radioactive Substances Detected in 4 Vegetables Imported from Japan," *Today*, March 25, 2011, p. 1.

41 Ken Belson and Hiroko Tabuchi, "Japan Confirms High Radiation in Spinach and Milk Near Nuclear Plant," *The New York Times*, March 19, 2011.

42 William J. Broad, "Scientists Project Path of Radiation Plume," *The New York Times*, March 16, 2011.

43 William J. Broad, "Radiation Plume Reaches U.S., But Is Said to Pose No Risk," *The New York Times*, March 18, 2011.

44 "Briefing Japan's Catastrophes: Nature Strikes Back" (2011).

45 Mitsuru Obe, "Radiation Threatens to Curb Relief Efforts," *Wall Street Journal*, March 16, 2011, p. 4.

46 Benjamin K. Sovacool, "A Critical Evaluation of Nuclear Power and Renewable Energy in Asia," *Journal of Contemporary Asia* 40(3) (August, 2010), pp. 369–400.

47 "Briefing on Nuclear Power: When the Steam Clears," *The Economist*, March 26, 2011, pp. 64–66.

48 "Briefing Japan's Catastrophes: Nature Strikes Back" (2011).

Index